Communicating Systems with UML 2

Communicating Systems with UML 2

Modeling and Analysis of Network Protocols

David Garduno Barrera
Michel Diaz

Library of Congress Cataloging-in-Publication Data

Garduno Barrera, David.
  Communicating systems with UML2 : modeling and analysis of network protocols / David Garduno Barrera, Michel Diaz.
     p. cm.
  Includes bibliographical references and index.
  ISBN 978-1-84821-299-2 (hardback)
  1. Computer network protocols. 2. Telecommunication systems. 3. UML (Computer science) I. Diaz, Michel, 1945- II. Title.
  TK5105.55.G37 2011
  004.6--dc23

                                                                                    2011014706

British Library Cataloguing-in-Publication Data
A CIP record for this book is available from the British Library
ISBN 978-1-84821-299-2

Printed and bound in Great Britain by CPI Antony Rowe, Chippenham and Eastbourne.

MIX
Paper from
responsible sources
FSC® C013604

Table of Contents

Preface

Advanced distributed architectures such as mobile embedded systems are becoming more and more complex but always need to keep the same quality. The design of such systems requires a full understanding of their internal behaviors and global interactions, and a mastery of how to represent, analyze and validate them, as well as an understanding of the accuracy of communicating entities and protocols.

Current books on network protocols give a basic knowledge of communication systems and present existing protocols layer by layer, but without explaining how to fully validate the communication mechanisms between them. A few books propose the use of a formal model to design protocols but they do not explain how these models can be created. Furthermore, these formal methods are not always fully adapted to modern high level programming languages, based on the object-oriented paradigm. The design software engineering approach required needs to integrate an object-oriented methodology and a protocol description language. Fortunately, both of these requirements are now provided by the Unified Modeling Language, or UML.

The UML language is composed of a simple and clear set of diagrams allowing a designer to easily model protocol mechanisms and architectures. UML is currently supported by an increasing set of description and analysis methods and tools.

This book not only presents a set of diagrams but helps the reader first to discover which diagrams to use for what kind of problem, and second, how to combine those diagrams into a coherent and meaningful model. As a consequence, the first goal of this book is to support readers interested in network protocols to first model their own requirements, describe their particular problem and, finally, build a full model of the protocol designed, together with its surrounding context.

Secondly, this book aims to show analysts or designers how to organize their modeling process into a logical and coherent set of steps, through an iterative and incremental approach to modeling and validating network protocols.

Using illustrated examples, this volume starts by modeling a few very simple communication cases whose purpose is to familiarize readers with the application of the object-oriented paradigm and state machine descriptions to network protocols.

Then, a chat application, a non-reliable medium entity and a transport protocol are successively modeled and tested by simulation. The purpose of these examples is to clearly illustrate how the different adjacent layers of a communication system are represented, designed and validated using a methodology based on the UML language.

Finally, the three adjacent layers, i.e. the protocol (under study) layer, the underlying (medium) layer, and the user (application) layer, are interconnected and linked together into a global coherent model. Again, this full system is represented by using a UML model and tested by the simulation of this model using existing UML tools.

By highlighting the design methodology, capabilities and properties of the models, we hope that this book will greatly help designers to understand, build and validate new, advanced distributed communicating systems, or indeed any other kind of systems.

David GARDUNO BARRERA and Michel DIAZ
May 2011

Chapter 1

Why Use UML to Model Network Protocols?

1.1. Modeling network protocols

1.1.1. *The complexity of communication protocols*

In the future, as advanced architectures such as multimedia systems, control systems, distributed embedded systems, etc, become increasingly complex, they will still need to fulfill a set of well-defined requirements regarding their global quality. Meeting these requirements needs a deep understanding of the full distributed system, i.e. of all its local entities and all the communications between those local entities, in order to be able to represent and anticipate their resulting global behavior.

As a consequence, a system's (global) behavior will depend on:

– local activities and their local data; and also on

– the messages that are sent, received and processed by the various interconnected entities.

A communication system with all its communication capabilities and configurations is often the basic and most difficult sub-system to correctly understand and build. The reason why this particular problem can be very complex, and lead to many design problems and implementation bugs, needs to be emphasized. To understand it, consider N processors that are connected: they can communicate, at a given instant, 2 by 2, 3 by 3, etc, or use full communication between them all, in which the N processors interact. The sum of the resulting combinations, which may need to be fully analyzed in order to understand the globally interconnected behavior, is then of the order of two to the power N.

From this number, it is easy to understand the potential complexity of the resulting conceptual problems, and in particular the increased difficulty which arises when going from interconnecting a few processors to interconnecting a large number of them: when N increases from 2 to 10, the number of potential interactions grows from 4 to more than 1,000.

It should therefore be clear that designing such distributed architectures can lead to a very complex conceptual design task, which, as a consequence, has to be based on a well-defined methodology in order to manage all the potential difficulties arising when building the system.

To manage such complex computer systems, a designer will usually at least go through the four following steps:

– specification: defining the system requirements, i.e. expressing the required functions and properties in a clear, complete and unambiguous manner;

– architecture: proposing the conceptual organization of the selected elements, composing the system in order to fulfill the requirements requested with the best performance possible;

– implementation: realizing the functions in the real and fully deployed architecture;

– validation: checking that the system fulfills its requirements, i.e. that the implementation satisfies the required properties.

Of course, designers, whenever possible, should use appropriate support tools to make their tasks more efficient. For instance, a designer can use support tools to capture requirements, to structure a design, to derive the implementation and to conduct the validation, either by verification, by simulation or by testing.

More precisely, when a system is complex, after defining the required properties, the various processes and entities, and defining the way they communicate, it is necessary to analyze the accuracy of the design. Of course, such validation has to be done as soon as possible, since correcting errors is simpler when the system is in an early stage of development, i.e. before it is fully implemented.

In any case, at whatever stage the validation is conducted, an adequate approach has to be used in order to represent global system behavior. Thus, the approach has to be defined to allow the designer to check and validate the appropriateness of the proposed behavior with respect to the system specification.

Then, precise specifications, i.e. a precise definition of the behavior expected, and a clear validation approach, i.e. one that is able to control the design process, are essential, or at the least, very valuable.

The first approach used to express the system specifications and validate the design is based on the use of a precise model of the involved mechanisms, functions and sub-systems, etc. Furthermore, the obtained semantic representation of the system must allow the designer to correctly express both the required properties and the required behavior of the designed system.

As a consequence, validating a proposed design becomes simple, since it entails checking the behavior of the designed system model against the required properties.

Note that, after a successful design validation phase based on such models, a system is considered "well designed", and can be implemented. Normally, one or more validation phases can be conducted if needed, at different stages of the design.

1.1.2. Traditional modeling

Specification approaches have been used for many years for the verification of communication systems and protocols.

Although they had no precise specification in the early years, the complexity of protocols and the reliability of ISDN[1] telecommunications systems have led to the development of formal specifications using formal descriptions. Two principal approaches were first proposed in order to write these specifications:

– basic formal models, such as finite automata, Petri Nets, process algebras, etc.; and

– formal description techniques, such as Estelle, LOTOS, and SDL.

Later, semi-formal models were developed and used to represent more complex systems, and to express many different specification and design perspectives.

---

1 Integrated Services Digital Network.

The basic reasons for using these models are:

– to represent, if possible in a non-ambiguous way, the requirements and specifications of the system to be designed;

– to translate the specification, using given rules and model simplification when needed, into a mathematical or executable model;

– to verify or analyze the mathematical or executable model, which might be highly complex, in order to validate the resulting model of the system and so the underlying designed system.

1.1.2.1. *Basic models and formal description techniques*

Basic models which do not include language-specific operators are the simplest solutions for representing mechanisms in a very abstract way. However, it should be noted that, as a consequence of their simplicity, such basic models may not describe all the details involved in the systems. Moreover, if these details are of importance for any resulting behavior, other more complete and complex models are needed.

It follows that each model has its own particular characteristics, which are more or less relevant for any specific design. Consequently, the choice of the model depends on the system under study and on the properties to be analyzed, as the model must be at the same time able to describe the properties and the design to be checked.

Two principal basic formal models are now presented: state automata and Petri Nets.

1.1.2.2. *Automata and state machine models*

The first models proposed for numerical systems led to the definition of finite automata, or state machines [BOO 67]. These automata or state machines are based on three fundamental assumptions, often given implicitly. The first two of these are:

– Assumption 1: The system has inputs and outputs, and there exist, for the systems under consideration, a concept of a "global state" and a set of these global states, and, furthermore, an explicit representation of these states that can be precisely defined, usually using the set of system variables.

– Assumption 2: The global behavior (an operational behavior) of the system can be expressed, starting with an initial (global) state, by a set of transitions that go from this initial state, the present state, to a set of other states, the next states. Note that there is potentially one next state per transition. Each transition goes from one state, the present state, to another (by definition) reachable state, the next state, and this repeatedly from all reachable states to all other reachable states. Note that if one reachable state is equal to one already existing state, then there is no new next state, and the end of the transition will be the already existing state.

More precisely, a transition between two states will take place when a given condition becomes true during the system evolution. A given condition can be, for instance, the occurrence of an enabling event (e.g. an entry whose value changes to 1), or the fact that a predicate becomes satisfied (true). A transition occurs at that point, or is fired when the system is in a present state and the condition associated with a transition leaving that present state becomes true.

The description of a transition (and so of the corresponding part of the system behavior) is represented in a model by drawing an arc starting from the present "before the condition" state and going to the next "following this condition" state.

As a consequence, the full behavior of the model is described by a global diagram, representing the way the system operates, giving the set of all possible reachable global states (represented by circles) and all possible transitions (represented by arcs) that exist between those states.

Note that this (behavioral) model can be built recursively, step by step, starting from one state, "the initial state", i.e. the well-defined specific state from which the behavior starts.

The first two assumptions are complemented by a third very important one in order to define (and then construct) the behavior:

– Assumption 3: the indivisibility of the transition between two states.

Indivisibility means that, when one condition occurs in a given state and triggers a transition between two states, the firing of the transition is indivisible, i.e. the transition is fully completed before entering the next state. Thus, this means that another triggering condition cannot be taken into account before the next state is reached. So, either such a triggering event should not occur or, if it does, it should not be processed, and its processing and its effects are delayed until the system is in the next state.

This means that there is no state between the present state and the next state, as the system leaves the present state and reaches the next state "indivisibly" (i.e. it reaches the state defined as "next state" for this transition). Of course indivisibility has to be satisfied whatever is done in the system during the transition, e.g. whatever the outputs produced.

In their basic definition, automata and state machines communicate with their environment by inputs and outputs.

In particular, the transitions between two states are defined by the values of the inputs, and each input is potentially able to trigger or enable one transition.

Outputs are defined either in a state (called a Moore machine) or a transition (called a Mealy machine). The former is defined by pairs of states/outputs, and the latter by pairs of arcs/outputs. In the specific case of the initial state, if the outputs are defined by states, then they must be defined by the initial state, and be the initial values of the system outputs.

In particular, the assumption of indivisibility for state machines implies:

– first, that a transition is executed when the significant condition of the automata enables the transition; and

– second, the next state has to be reached before a new condition can enable a new transition in this new, next state.

Indivisibility means that a transition and its outputs (the actions in that transition) must be completed before reaching the next state.

Thus, a global behavior of the model can be defined by considering, consecutively, the set of the states, inputs, transitions and outputs that define the system execution.

Furthermore, note that the assumption of indivisibility implies that, when complex actions (e.g. producing many outputs or running computations) are associated with a transition, those actions must be completed before reaching (and thus defining) the next state.

As a consequence, whatever the transition actions and outputs, only two global states exist:

– one before the transition, i.e. the starting global state, containing all the values existing at the instant before the transition is executed; and

– one after the execution of the transition, the next (global) state, containing the new values of the automata or state machine, i.e. the values either left unchanged by the transition, or modified by the complete execution of all actions associated with the transition.

Of course, for the model to correctly represent the real behavior of the system, the behavior of the system under study must also satisfy the assumption of indivisibility, i.e. the behavior of the real system must satisfy the indivisibility assumption (to be coherent with the behavior of the model).

As a consequence, the model must represent the real indivisibility that exists in the modeled system. Conversely, if some sub-behaviors must be considered as divisible (i.e. they are not indivisible), then they cannot be represented by a single transition and must be represented by a set of transitions, each of which represent the required indivisible sub-behaviors.

The model of a state machine, a state diagram, can be represented by a directed graph defined by the following tuple (S, I, O, δ, q_0):

– S: a finite set of vertices (the states);

– I: a finite set of inputs;

– O: a finite set of outputs;

– δ: a state transition function, describing the transitions between states $I \times S \rightarrow S$;

– q_0: the initial state.

The output function, for a Mealy machine, maps the input and state sets onto the output set, i.e.: $I \times S \rightarrow O$.

Sometimes, a set of final states is also defined.

1.1.2.2.1. Models of tasks and processes

Of course, a program or a process can be represented by a state machine:

– the initial state is given by the values of the program counter and the program variables, right after program initialization;

– the triggering or execution of a transition is defined by its set of actions, i.e. the executions of the program instructions associated with the transition;

– each transition execution leads to a next state that includes the new values of the program counter and the values of all program variables (i.e. the new values of the program variables that have been modified by the transition, and the old values of the program variables that have not been modified by the transition).

Again, any divisible behavior of the system must be broken up into indivisible sub-behaviors, and each of these indivisible sub-behaviors can be represented by a transition.

In particular, if some actions can be suspended, then their models must be selected accordingly. Of course, to do this, the designer must correctly map between the model and what it represents in the real system.

1.1.2.2.2. A few simple models

Let us consider the two machines given in Figure 1.1(a), composed of circles, states and arcs (the transitions between the states). Note that each transition has one, and only one, starting state, and one (and only one) next state.

Let us suppose that, to mark the initial state, at the initial instant, a token is drawn in the state (a black filled circle in the middle of the state). Note that the initial state can also be marked by an arrow entering the state that is the initial state, and coming from no other state (as in Figure 2.10 in the next chapter).

When a transition is executed, after being enabled by an event, the fact of going from the present state to the next state is called the execution, or the firing, of a transition, and this firing can be represented graphically by passing the token from the present (starting) state to the next (following) state. Note that, for a transition to be fireable, the token must be at the "input of the transition", on the side opposite to the arrow.

One token always exists in the graph and indicates the present state of the behavior, and each state represents a global state of the model.

In Figure 1.1, in the notation "A/B", "A" is the input that triggers the transition when it happens or is true, and "B" is the output produced by the firing of that transition.

From what we have already described, this means that, when the token is in the input state of the transition, the arrival of the input event "A" causes the execution of the transition from the present state to the next state, and the firing of that transition produces the output action "B".

The partial machine in Figure 1.1(a)(top) represents a state having a set of next states, each one being enabled by an input e_i (for instance, where e_i appears or when e_i becomes "true"). The firing of the transition produces an output o_i. Each input leads to a next state. Furthermore, the inputs must be exclusive: only one of them can be true at a given instant of time, leading to one next state.

The machine in Figure 1.1(a) (bottom) gives a very simple automata, having three states, with one input (0 or 1), and two outputs, also binary (0 or 1). In the initial state, there are two possible inputs:

− 1, which leaves the automata in the same state (as the next state is the same); and

− 0, which triggers the transition 0 / 00.

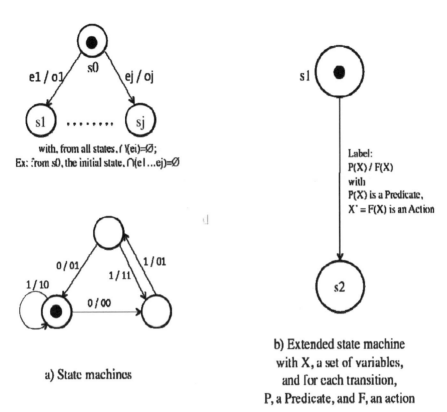

with, from all states, ( )(ei)=∅;
Ex: from s0, the initial state, ∩(e1 ...ej)=∅

a) State machines

Label:
P(X) / F(X)
with
P(X) is a Predicate,
X' = F(X) is an Action

b) Extended state machine
with X, a set of variables,
and for each transition,
P, a Predicate, and F, an action

Figure 1.1. *Simple and extended state machines. (a) Partial (top) and full (bottom)*
state machines; (b) extended state machine

Note that, as a consequence, while the input is "1" the state and the outputs do
not change, but when it becomes "0", the transition is fired and the produced output
is "00".

Note also that in one state, on the right-hand side in Figure 1.1(a) (bottom), the
transition enabled by (the input value) "0" is not given, i.e. it is not specified, and in
such cases, to simplify the drawing, the next state for that value is the same state
(and the loop for that state in not given).

Figure 1.1(b) models a complex transition of an extended state machine, being,
for example, a step in a program, with a logical condition. Such a model has the
following behavior:

– when s1 is marked, if Predicate P(X) is false, then the transition is not enabled
and cannot fire, and the system stays in the same state;

– when s1 is marked, if predicate P(X) is or becomes true, then the transition concerned is enabled; the transition is then fired and the program progresses towards its next state, state s2, and this firing runs the output action, i.e. this firing executes the procedure F(X), producing X', the new set of program variables (the new data values). Note that P represents the enabling condition of the evolution (of firing) and F gives the action(s) performed on the program variables (new values) when firing the transition.

1.1.2.3. *Petri Net models*

Petri Net (PN) models and their extensions generalize state machines and are of high interest, as having provided the first approach and semantics for concurrent systems they can be used to model the behaviors of basic parallel and distributed mechanisms. Furthermore, they:

– define an easy graphical support for the representation and understanding of basic parallel and distributed mechanisms and their behaviors;

– provide an easy to understand extension to state machines and automata (discussed above);

– express, in a simple manner, the main basic concepts needed in communications, including waiting and synchronization, and, furthermore, can take into account their temporal and stochastic aspects;

– can be used, at the same time, first for the definition of models, and then for the validation of the behaviors of those models, by allowing the definition of analysis methods and software tools;

– ensure, being unrelated to a particular programming language, the independence of a specification with respect to its implementation.

Generally speaking, Petri Nets are defined as extensions of the concept of transitions in state machines: transitions in PNs can have not one but several input (ingoing) arcs at the same time, and several output (outgoing) arcs.

Note that, in such a case, the circles do not represent global states: they represent local states and are called the PN's "places". The tokens in the places also represent local information. The set of all tokens in all places is called the PN's "marking". It follows that a set of tokens is the "state" of the PN. Finally, the set of all markings is the set of all reachable states of the PN.

Such a simple extension proves to be particularly powerful to express and analyze parallel and distributed behaviors, since:

– several tokens can exist in the model at the same time (and not just one), meaning that several independent parallel behaviors can be represented at the same time in the model;

– there are no explicit global states, and the global state of the PN is the set of all places and tokens.

Furthermore, many PN-based models exist, forming a family of PNs, starting from simple traditional state machines, and moving up to models allowing system designers to handle functional (qualitative) and non-functional (e.g. quantitative) behaviors in an integrated way, i.e. both the temporal behaviors (Time PNs) and stochastic behaviors (Stochastic PNs) of systems.

The scope of this book does not include a detailed explanation of PN models. Interested readers will find more information and examples in [DIA 09]. Let us note, however, that UML uses the concepts of "join" and "fork", which are concepts drawn from PN models. These simple concepts will be explained through examples in Chapters 2 and 5.

1.1.2.4. *Formal and semi-formal languages*

The basic models we have discussed so far cannot represent all possible actions and mechanisms that are used in complex communication protocols, as such protocols have to handle data, predicates and actions, and even, in a general sense, can manipulate the software architecture of a system.

The first formal approaches to represent the complete behavior of protocols were SDL (Specification and Description Language) [SDL 09], Estelle and LOTOS [VAN 89]. The description of the control routines in SDL and Estelle is based on state machines, while LOTOS is based on CCS (Calculus of Communicating Systems). Data management in Estelle is based on the Pascal language, while LOTOS is based on abstract data types.

We should emphasize that, as SDL and Estelle are based on state machines, it is quite easy to represent program behaviors, as discussed earlier, but it is also quite easy to describe communicating mechanisms and protocols. Finally, the concept of "next states" also provides an easy way to manipulate the corresponding behavioral representations of the models.

Recent versions of UML include some SDL features, mainly for state machine diagrams and activity diagrams. In the following chapters, we will use these diagrams in order to model the behavior of the communicating entities under study.

1.1.2.5. *Towards a new modeling language*

The disadvantages of these traditional modeling methods, models and languages are that they were very specialized and focused on protocols. This has also led to the definition of very specialized tools, and to the need to acquire specific knowledge. Thus, they were aimed at a very particular and specialized audience.

These difficulties led to the definition and use of a more general approach, and so to the design of general languages and tools, able to be used by far more people (which justifies the costs needed to develop and maintain the language and associated tools), in a user-friendly and easy to understand manner.

Of course, the properties coming from the use of such a description approach, including the validation of the model, are still more necessary because, as the protocols become more complex, the possibility of making protocol errors increases.

1.1.3. *Traditional validation*

The need to master the design of complex communicating systems led to the development of a validation methodology whose aim was to validate the design as soon as possible during the software lifecycle. In particular, it encouraged the validation process to be started during the specification phase.

The purpose of such an early validation is to detect errors as soon as possible, before going into the details of the implementation. This is because a specification or design error becomes much more difficult to detect, and much more difficult and expensive to correct, when the system is fully designed, and still more so when it is implemented.

As a consequence, formal models, and to a lesser extension semi-formal models, must always be able to:

– describe the semantics of local behaviors and communications;

– describe global expected behaviors;

– validate the model behaviors obtained during the design stages; and

– be able to lead to a set of supporting software tools.

In general, designers understand the fundamental semantics of a system, i.e. the basic building blocks, quite well, but cannot be sure of really understanding the system's global behavior, particularly with regard to rare functional issues, as in the case of quite unusual usage scenarios.

The validation phase tries to check the correctness of a model's behavior for the designed system.

As discussed previously, the most frequently used approaches for "reachability" or accessibility analysis are based on building and manipulating a (system) accessibility or reachability graph, i.e. a graph representing all possible behaviors of the model.

Verification analyzes the structure of the graph of all reachable (or accessible) states, i.e. the set of states that are reachable or accessible from a given initial state. Remember, for example, that the set of these reachable states can be computed, for state automata, by firing all transitions from all possible states, i.e. for all possible states starting from the initial state.

In some cases, for instance for extensions of PNs, this graph allows designers to check not only non-temporal behaviors but also temporal behaviors, i.e. the behavior that depends on explicit values of time, as well as stochastic behaviors, i.e. behavior that depends on explicit values of distributions.

The problem is that, for complex systems, checking models' properties by using algorithms applied to the accessibility graph leads to the problem of combinatorial explosion.

This means that, when there are too many states, it becomes too expensive or too time consuming to have them all checked, and a simpler solution is needed.

The idea is then not to consider all possible behavioral sequences, but to define a dedicated set of these, i.e. a set of sequences that are likely to show potential errors.

Note that such an approach defines, in fact, a simulation approach.

The only difference with a classical running simulation is that, when using formal or semi-formal models, these models can be used to define, in a clear and very often clever way, the set of inputs that have to be used to define the selected test sequences, leading as a consequence to the definition of more intelligent model based simulations.

UML based simulations will be extensively used and explained in this book.

1.1.4. *Need for a unified language for description, validation and simulation*

In parallel to communication protocols, software designers started to represent software architectures and systems by using object-oriented approaches and models

[BOO 94]. UML became the main language used in this area. Later, it became clear that, in complex systems, objects have to cooperate, and that the resulting communication sub-system must handle protocols. This means that we need at the same time to include an explicit representation of an object's behavior and an explicit representation of the communication between objects.

As a consequence, this led to the definition of an approach that is able to integrate both aspects in the same full model, i.e.:

– first, the object-oriented model to represent the architecture of the objects; and

– second, the state machine approach to represent the behavior of the objects communicating.

We selected this integrated approach, based on UML and SDL, for this book because it proved to be quite powerful in order to design and validate specific and general communication protocols and architectures.

Different tools have been developed in order to support the design of complex systems by using UML. These tools help the designer first to produce the UML description itself. Then, based on the state machine concepts, they also help users to conduct the simulations needed in order to validate the described behavior, and so the underlying system.

As a consequence, protocol or system designers are greatly supported in writing UML models for systems, and in simulating models, by the use of the different simulation tools which have been developed. We will see that it can be very efficient to conduct a model and system definition, and a model and system validation, by using a progressive approach.

Finally, note that simulation tools only provide a simulation engine, and generally do not provide the test to be applied during simulation. This means that the definition of the simulation sequences, i.e. the test sequence applied to a model, is left to the protocol or system designers, or to dedicated protocol or system testers.

1.2. UML as a common language

1.2.1. Overview

Throughout history, human beings have created models to represent complex systems. These models help to reduce risks, to better understand problems, to control complexity, and to communicate ideas. This communication, and in particular that between the members of a team, is achieved by using visual artifacts, decomposing the problem and creating specialized views, each view representing a different degree of complexity from the point of view of the different members of the team.

Within the software context, this is particularly true since modern companies work with worldwide distributed teams, and systems become more and more complex and diversified.

The motivation to create a model of a system under study is also to manage the variety of different experts involved on a software system: users, engineers, analysts, designers, developers, vendors, technicians, quality team, trainers, etc. This is because each group has a different view and different knowledge of the system, and is using a different language to describe it.

Under these conditions, the necessity of a unified communication language is undeniable.

UML is a modeling language defined for the software context in order to describe and analyze requirements, to describe and analyze any kind of system, to design and model solutions, to communicate between the members of a team, etc.

UML is well suited to any kind of system, domain or process. Within the telecommunications framework, UML proposes certain features which perfectly implement (or replace) the existing methods, models and languages traditionally used for this discipline.

1.2.2. *The beginning*

Historically, UML started as a group of separate methods aiming at modeling software systems with the new Object-oriented paradigm. Among all these methods, three of them had real success: Booch, created by Grady Booch; Object Modeling Technique (OMT), created by James Rumbaugh; and Object-oriented Software Engineering (OOSE), created by Ivar Jacobson. These three methods were fused in 1995 and created the first version of the Unified Modeling Language (UML).

In 1997, UML was submitted to the Object Management Group (OMG). At the end of the same year, the OMG approved a new revision (UML 1.1) and adopted UML as its official modeling language.

Since then, UML has increased its success and has increased its power of expression, mainly with the active participation of the main actors in the industry.

In 2000, a main revision was approved by the OMG and called UML 2.0. This new revision was composed of 13 diagrams (instead of 9 in version 1.4).

The last official revision, UML 2.3, defines 14 diagram types and was released in May 2010. This revision is divided into two major diagram types: structure diagrams and behavior diagrams, as shown in Figure 1.2.

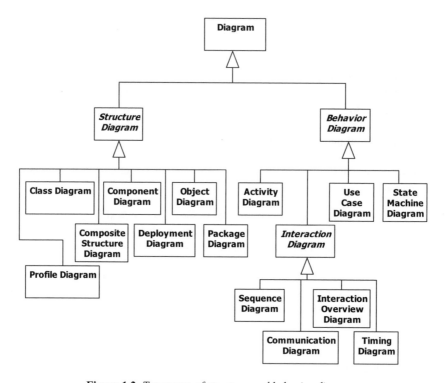

Figure 1.2. *Taxonomy of structure and behavior diagrams*

Structure diagrams allow us to describe the static structure of the objects in a system; these diagrams take no account of time and represent the meaningful concepts of an application.

Behavior diagrams show a methods' behavior, collaborations, activities and the state histories of a system; in other words, the dynamic behavior of the system.

1.2.3. *Brief review*

1.2.3.1. *Class diagram*

One of the main diagrams in UML, at least from the point of view of object-oriented programming, is the class diagram. This diagram represents the main concepts (or objects) and their interactions. It is used by most of the modeling tools in order to automatically generate the running program code of a system.

The following nodes and edges are typically drawn in a class diagram:

– Association;

– Aggregation;

– Class;

– Composition;

– Dependency;

– Generalization;

– Interface;

– InterfaceRealization;

– Realization.

Figure 1.3 shows an example of a class diagram extracted from the current UML specification [SSU 10].

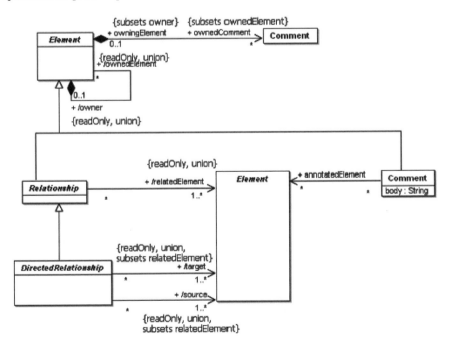

Figure 1.3. *Example of a class diagram*

1.2.3.2. *Package diagram*

Package diagrams show the logical architecture or organization of the system under study. They describe the dependencies and relationships between the different logical blocks composing a system.

The following nodes and edges are typically drawn in a package diagram:

– Dependency;

– Package;

– PackageExtension;

– PackageImport.

Figure 1.4 shows a package diagram extracted from the current UML specification [SSU 10].

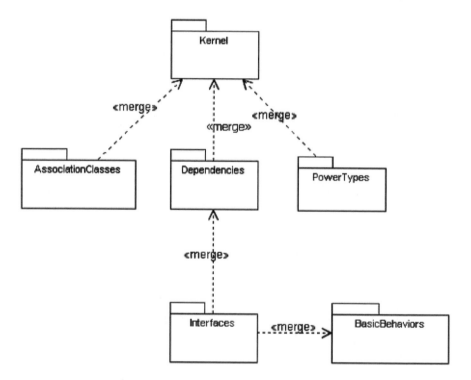

Figure 1.4. *Example of a package diagram*

1.2.3.3. *Object diagram*

An object diagram depicts a complete or partial view of the instances of the elements composing the system at a specific time. It focuses on instances and attributes values.

The following nodes and edges are typically drawn in an object diagram:

– Instance Specification;

– Link (i.e. Association).

1.2.3.4. *Component diagram*

A component diagram shows how complex structures (components) are linked together in order to compose software systems. It also describes the blocks' interfaces, both those provided and those required.

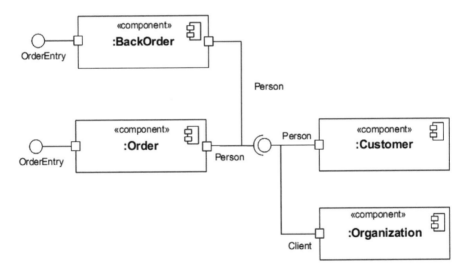

Figure 1.5. *Example of a component diagram*

The following nodes and edges are typically drawn in a component diagram:

– Component;

– Interface;

– Component Realization, Interface Realization, Usage Dependencies;

– Class;

– Artifact;

– Port.

Figure 1.5 shows an example of a component diagram extracted from the current UML specification [SSU 10].

1.2.3.5. *Composite structure diagram*

A composite structure diagram shows the internal structure and organization of a class.

Figure 1.6 shows an example of a composite structure diagram extracted from the current UML specification [SSU 10].

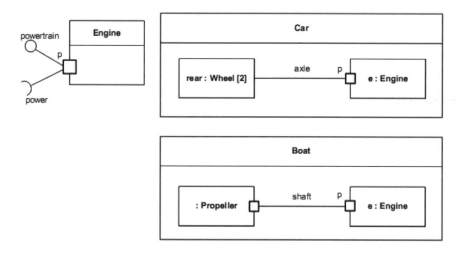

Figure 1.6. *Example of a composite structure diagram*

1.2.3.6. *Deployment diagram*

A deployment diagram defines the execution architecture of a system, which represents the assignment of software artifacts to nodes. It describes the hardware used in the deployment of a system together with the execution environment and the artifacts deployed on that hardware.

Figure 1.7 shows an example of a deployment diagram extracted from the current UML specification [SSU 10].

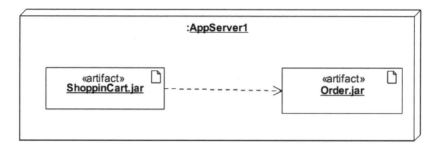

Figure 1.7. *Example of a deployment diagram*

1.2.3.7. *Sequence diagram*

A sequence diagram describes the set of messages exchanged between participants in chronological order. It is sometimes called an event diagram or event scenario.

Figure 1.8 shows an example of a sequence diagram extracted from the current UML specification [SSU 10].

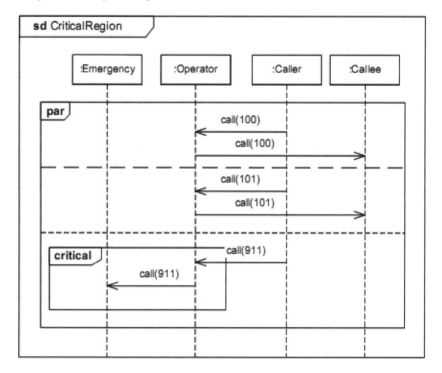

Figure 1.8. *Example of a sequence diagram*

1.2.3.8. *Profile diagram*

A profile diagram is the latest diagram being added to a specification. It describes the stereotypes included in the metamodel as classes. It includes the ability to adapt existing metamodels in order to extend them with the intention of adapting them for different purposes, such as in the case of J2EE, .NET, real-time or business process modeling.

Figure 1.9 shows an example of a profile diagram extracted from the current UML specification [SSU 10].

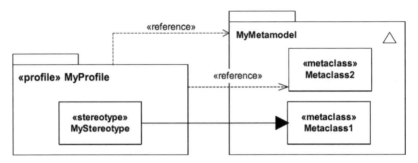

Figure 1.9. *Example of a profile diagram*

1.2.3.9. *Activity diagram*

An activity diagram describes a system's or component's behavior as a workflow representation. It can be also be used in order to describe the business and operational behaviors of an overall process.

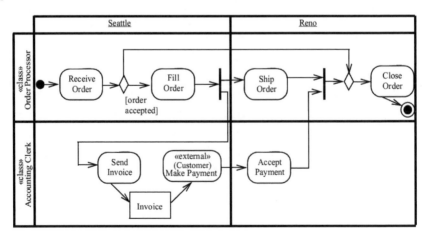

Figure 1.10. *Example of an activity diagram*

Figure 1.10 shows an example of an activity diagram extracted from the current UML specification [SSU 10].

1.2.3.10. *Communication diagram*

A communication diagram describes the interactions between participants as an ordered sequence of messages. The sequencing of the messages is given through a sequence numbering scheme.

Figure 1.11 shows an example of a communication diagram extracted from the current UML specification [SSU 10].

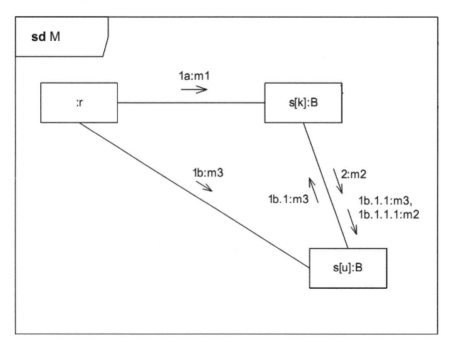

Figure 1.11. *Example of a communication diagram*

1.2.3.11. *Interaction overview diagram*

Interaction overview diagrams focus on an overview of the flow of control, where the nodes are other interaction diagrams. This type of diagram is very useful in order to describe very complex scenarios by including or referring to other diagrams.

Figure 1.12 shows an example of an interaction overview diagram extracted from the current UML specification [SSU 10].

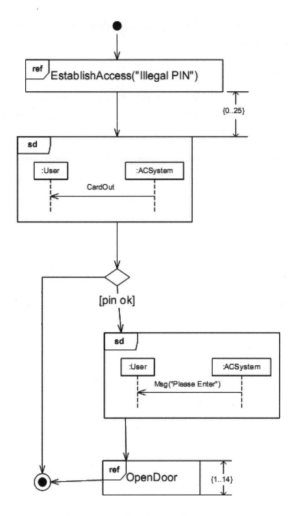

Figure 1.12. *Example of an interaction overview diagram*

1.2.3.12. *Timing diagram*

Timing diagrams describe the behavior of a single class, or the interactions between many of them, focusing on time constraints over a given duration. They describe the occurrence of the triggering events, which cause changes in class conditions or states.

Figure 1.13 shows an example of a timing diagram extracted from the current UML specification [SSU 10].

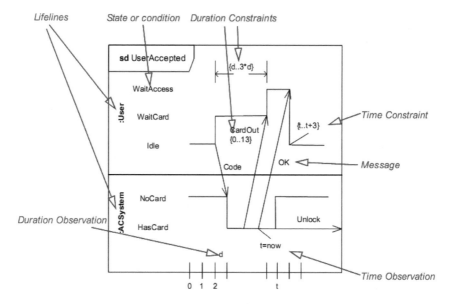

Figure 1.13. *Example of a time diagram*

1.2.3.13. *Use case diagram*

Use case diagrams describe the different usages of a system from the point of view of its external actors. They are particularly useful for capturing and analyzing requirements and organizing the overall software lifecycle.

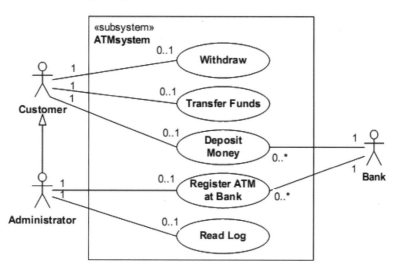

Figure 1.14. *Example of a use case diagram*

Figure 1.14 shows an example of a use case diagram extracted from the current UML specification [SSU 10].

1.2.3.14. *State machine diagram*

State machine diagrams are a highly improved realization of the well-known (finite) automata. A state machine diagram describes the behavior of a single class or the overall system as a set of states and transitions. This diagram combines elements coming from different mathematical concepts used in computer science and which were explained at the beginning of this chapter, such as automata, Petri Nets and SDL.

Figure 1.15 shows an example of a state machine diagram extracted from the current UML specification [SSU 10].

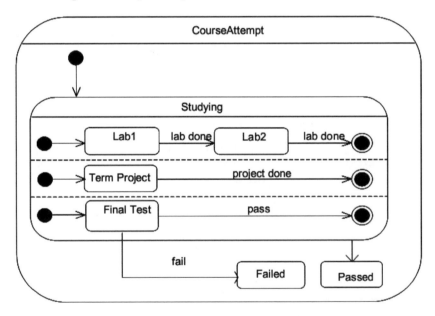

Figure 1.15. *Example of a state machine diagram*

1.2.4. *UML for network protocols*

As we have already said, UML is well suited for any kind of systems. For network protocols, UML owns all the advantages of a mathematical tool or language, with the huge advantage of being object-oriented. This last characteristic provides numerous benefits. An object-oriented model is perfectly understandable by any developer, it facilitates the work of the simulation tools, and facilitates the creation of different simulation tools. Finally, an object-oriented model can be enhanced with blocks of programming code, such as C++ or ADA.

In addition, the usage of use case diagrams may significantly facilitate the writing and analysis of requirement specifications, by dividing the main system into functional independent scenarios. This division may lead to a functional and logical separation into different software modules which can then be modeled, implemented and tested separately.

Sequence diagrams might be used together with use case diagrams in order to better explain the latter. Moreover, the use of time and duration constraints, interaction use (called "references" in UML) and combined fragments (called "control blocks" in UML) add an enormous power of expression to these diagrams.

The current definition of state machine diagrams allows the designer to express almost any complex behavior. Indeed, traditional state machines [HAR 87] were extended with the use of composite states, explicit action nodes, choice pseudo-states, history and deep-history pseudo-states, exit points and parallel regions. This new version of state machines is capable of representing the behavior of separate modules, internal methods and the behavior of the overall system. Finally, thanks to code translators, these state machines are easily mapped into a running code, and executed and traced by many Computer-Aided Software Engineering (CASE) tools, such as TAU G2, Enterprise Architect and Rhapsody.

Composite structure diagrams and deployment diagrams are very useful in order to represent how the different elements of a distributed system might be interconnected and deployed over a hardware platform or inside a server.

1.2.5. Some general UML tools

Many different tools supporting the use of UML are currently available. Most of them are not free, such as TAU G2, Enterprise Architect and Rhapsody.

However, some free tools, such as Star UML, can be found. It is even possible to find web-based and open source tools. Every tool is likely to have its own capabilities and particularities. Nevertheless, since all of them use the same language (UML), most of the modeling work can be performed perfectly well using any of them.

Note that three tools have been used in this book: Star UML, an open source tool [STARUML]; Enterprise Architect, from Sparx System [ENT]; and TAU G2 [TAUG2].

Finally, remember that it is assumed in this book that readers are aware of UML and, more precisely, that they are aware of the main concepts of the language. If this is not the case, readers can refer, for example, to [SSU 10] and [DRU 06].

1.3. Chapter summary

This chapter has presented the main concepts involved in the specification and validation of communication systems and protocols. The main basic models for protocols have been described, together with their underlying basic semantics.

The chapter has also explained why and how UML and object-oriented modeling facilitate the description, analysis, design, simulation and testing during the design of a system, including its requirements, interactions and interfaces with external actors, the behaviors of internal components of the overall system, its logical architecture and its deployment strategy.

Finally, validation and simulation concepts have been provided.

The next chapters will present a step-by-step approach, supported by a corresponding set of examples of increasing complexity, together with related UML descriptions. The purpose of this is to help readers by giving them an increasingly greater understanding of how to analyze and validate real communicating systems.

1.4. Bibliography

[BOO 67] BOOTH T., *Sequential Machines and Automata Theory*, John Wiley & Sons, New York, USA, 1967.

[BOO 94] BOOCH G., *Object-Oriented Analysis and Design with Applications*, Benjamin Cummings, Santa Clara, CA, USA, 1994.

[DIA 82] DIAZ M., "Modeling and analysis of communication and co-operation protocols using Petri Net based models", Tutorial Paper, *Networks Computer*, December 1982.

[DIA 89a] DIAZ M., VISSERS C., "SEDOS: Designing Open Distributed Systems", *IEEE Software Magazine*, Vol.6, No.6, pp.24-33, November 1989.

[DIA 89b] DIAZ M., ANSART J.P., COURTIAT J.P., AZEMA P., CHARI V. (eds), *The formal description technique ESTELLE. Results of the ESPRIT/SEDOS project*, (SEDOS Software Environment for the Design of Open distributed Systems) North Holland, 1989.

[DIA 09] DIAZ M. (ed.), *Petri Nets: Fundamental Models, Verification and Applications*, ISTE Ltd, London and John Wiley and Sons, New York, 2009.

[DRU 06] DRUSINSKI D., *Modelling and Verification Using UML Statecharts*, Elsevier, 2006.

[ENT] Enterprise Architect, http://www.sparxsystems.com/.

[GAR 05] GARDUNO BARRERA D., A differentiated quality of service oriented multimedia multicast protocol, PhD Thesis, 2005.

[HAR 87] HAREL D., "StateCharts: A visual formalism for complex systems", *Science of Computer Programming*, 8(3):231–274, June 1987.

[MUR 89] MURATA T., "Petri Nets", *Proceedings of the IEEE*, Vol.77, No.14, April 1989, p. 541-579.

[SDL 09] SDL Forum Society, http://www.sdl-forum.org/, 2009.

[SSU 10] Unified Modeling Language™ (UML®), Superstructure specification, 2010-05-05, http://www.omg.org/spec/UML/2.3/

[STA] Star UML, http://staruml.sourceforge.net/en/

[TAU] IBM, "Rational Tau", http://www-01.ibm.com/software/awdtools/tau/

[VAN 89] VAN EIJK P.H.J., VISSERS C.A., DIAZ M. (eds), *The Formal Description Technique LOTOS. Results of the ESPRIT/SEDOS Project*, North Holland, 1989.

Chapter 2

Simple Transmission

2.1. Introduction

The examples presented in this chapter are intended to introduce the key features of UML that need to be used for protocol modeling. We start with a simple protocol, implementing an "echo" to a message received. We then add more and more features to this protocol in order to obtain a simplified full-duplex data sending protocol.

2.2. Echo

2.2.1. *Requirement specification*

This first example aims at modeling a simple protocol with the following characteristics:

– there are two communicating entities: client and server;

– the client sends a "Hello_req" message to the server, and the server answers with a "Hello_res" message;

– system ends.

Let us represent these requirements by using a communication diagram (see Figure 2.1).

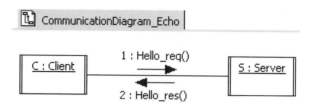

Figure 2.1. *A communication diagram for the Echo protocol*

2.2.2. *Analysis*

2.2.2.1. *Sequence diagram*

First, let us create a simple sequence diagram representing the expected behavior in a chronological order. For this, we will use a sequence diagram[1].

First, the client sends a Hello_req message to the server. On receipt, the server answers with a Hello_res message. The system then finishes. Figure 2.2 represents this expected behavior.

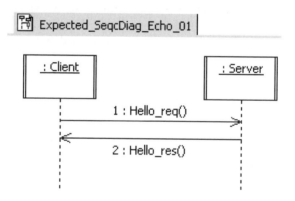

Figure 2.2. *Sequence diagram for the Echo example*

2.2.2.2. *Concerned classes*

Now, let us represent each of the two network nodes as a single class. This first global class diagram (Figure 2.3) is very simple since it only contains two classes: client and server.

---

1 For more details on sequence diagrams, see Chapter 14, "Interactions", in [SSU 10].

Figure 2.3. *Class diagram for the Echo protocol*

2.2.2.3. *Signals list definition*

The next step consists in modeling the messages exchanged between the network nodes[2]. A good solution for gathering the messages going from one node to another is to use interfaces.

We create an Interface called "Client_2_Server" in order to group the messages sent from the client to the server, and an interface called "Server_2_Client" to group the messages sent by the server to the client. We then include a Message[3] named "Hello_req" in the first interface and a message named "Hello_res" in the second interface. The class diagram in Figure 2.4 shows these interfaces and their messages.

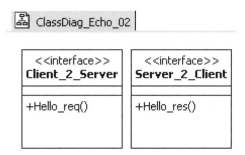

Figure 2.4. *Interfaces for the Echo protocol*

In the context of network protocols, nodes communicate through ports. So, we can associate a port to the client class and name it port P_C; then, we create a port called P_S for the server class. Figure 2.5 shows a class diagram containing these classes and their ports.

---

2 A natural way of discovering the exchanged messages is by modeling the functional scenarios through sequence diagrams.
3 A message represents an operation defined by a given class. This operation can be called from outside the class and might contain parameters. This concept is very similar to the signals used when modeling protocols: a message is sent by an entity and received by another. The reception of a message is intended to start an action in the receiver.

Figure 2.5. *Client and server classes with associated ports*

Now, let us associate the previously created interfaces with the client and the server classes.

A "provided" interface represents the set of public features or obligations that a given class offers or provides to the external world; in other words, it represents the set of operations that an external class can call on this given class. The set of interfaces "realized" by a class are its provided interfaces. The interface realization relationship from a class to an interface is shown in UML by representing the interface by a circle or ball, labeled with the name of the interface, attached by a solid line to the class or port that realizes that interface.

A "required" interface represents the set of operations that a given class needs or requires to be provided by another class in order to establish a communication. Required interfaces, which are specified by a "dependency" relationship between a class and the corresponding interfaces, specify the services that a class needs in order to perform its function and fulfill its own obligations to its clients. The dependency from a class to an interface is shown by representing the interface by a half-circle or socket, labeled with the name of the interface, attached by a solid line to the class that requires that interface.

We can summarize these concepts as follows[4]:

– input messages are gathered into a "provided" interface. UML represents this as a "realization" relationship between the class and the interface concerned;

– output messages are gathered in a "required" interface. UML represents this as a "dependency" relationship between the class and the interface concerned.

From the point of view of the client, the Server_to_Client interface is "provided" since it represents the messages that the client can receive. In the context of

---

4 For more details on dependency and realization relationships, and on provided and required interfaces, see sections 7.3.2, Dependency, 7.3.45, Realization, and 7.3.24, Interface, in [SSU 10].

protocols, the "provided" interface might be seen as the input messages set; see Figure 2.6(a).

From the point of view of the client, the Client_to_Server interface is "required" since it represents the set of messages that the client asks its communication partners to provide. In the context of protocols, the "required" interfaces might be seen as the output messages set; see Figure 2.6(a).

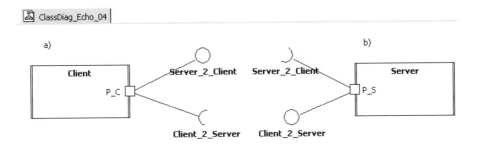

Figure 2.6. *Client and server classes with associated interfaces*

Conversely, from the server's point of view, Server_to_Client is the output interface while Client_to_Server is the input interface; see Figure 2.6(b).

Note that it is the port, and not the class itself, which is linked to the interfaces. Indeed, in the communications protocol domain, the nodes receive and send messages through the ports; these ports are therefore the artifacts that should be linked to the interfaces.

It is also possible to represent both classes together and the fact that they share the same interfaces, as shown in Figure 2.7.

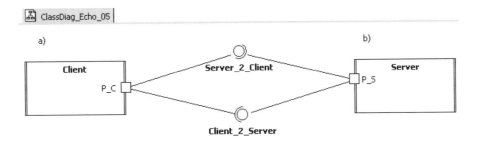

Figure 2.7. *Client and server classes sharing interfaces*

2.2.3. Architecture design

To represent the global system, we represent the client and the server classes as parts[5] of this System. We can define the association between these classes as a composition relationship, as shown in Figure 2.8.

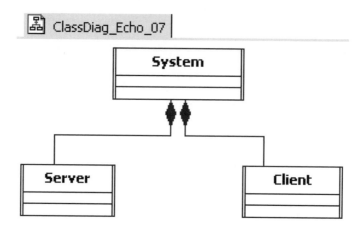

Figure 2.8. *Composition of system, server and client classes*

We also need a communication channel (Ch_C-S) between the client and the server. We could use a communication channel without any ports; however, in the telecommunications world, every communicating entity sends and receives data only through its ports. Thus, we link the communication channel to the corresponding ports in the client and server.

We can describe the communication path between the client and the server classes using a composite structure diagram (Figure 2.9).

At this point, we have added two parts to the system class: the server and the client. We have also defined that all the signals going out from the client class will be sent to the server class, and vice versa. The behavior of each part is described in the following section.

---

5 In UML, a part indicates a composition relationship; i.e. it indicates that an instance of a given class may contain a set of instances of other classes by composition. All such instances are destroyed when the containing class instance is destroyed. For more details on parts, see section 9.3.12, Property (from Internal Structures), in [SSU 10].

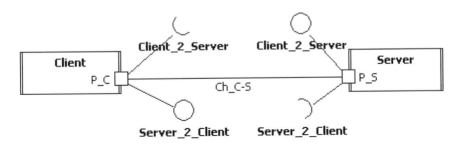

Figure 2.9. *Complete composite structure diagram*

2.2.4. *Detailed design: class behavior*

In the context of network protocols, we recommend the use of two types of diagrams to model the behavior of nodes: activity diagrams and state machine diagrams.

Activity diagrams are based on Specification and Description Language (SDL, [DOL 03]), a system description language standardized by the International Telecommunication Union (ITU-T) and well known in the telecommunications field.

It should also be noted that a few of the features of activity diagrams in UML are based on Petri Nets, such as the concepts of join and fork (seen later in section 2.4.5.4).

State machines, or Finite State Machines (FSM), extended by variables and actions, are the traditional way of representing the behavior of protocol entities: Figure 2.10 shows a simple example of such a finite state machine extracted from the current UML specification [SSU 10].

As usual, the change from one state to another through a transition is triggered by a stimulus. A transition may involve guard conditions, and it can execute actions and send output signals. Note that the expression [doorway ->isEmpty()] simply means that the close event will fire the related transition only if there is no object or person obstructing the doorway. This expression is defined in Object Constraint Language (OCL), the declarative language for describing rules that apply to (UML) models developed at IBM, and now part of the UML standard [OCL 10].

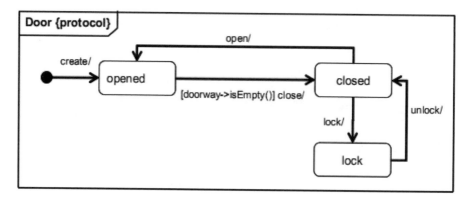

Figure 2.10. *Simple example of a state machine*

We can summarize transition notation as shown in Figure 2.11. In this figure, the system changes from state S1 to state S2 on receipt of InputSignalEvent only if the condition "Condition" is true. During the transition, the system executes the action "Action"[6]. Of course, every different modeling tool implements the standard in a slightly different manner. However, they all keep the same semantic meanings: conditions, input events and actions (or post-conditions).

Figure 2.11. *State machines: transition notation*

The good news is that the extended state machines defined in UML are quite general, and they allow us to use a very rich notation, which currently includes inputs, outputs, choices, time events, etc[7].

2.2.4.1. *Client class*

First, let us define the client's behavior. If we look at Figure 2.2 in section 2.2.2.1 above, we can see that, when starting and without any stimuli, the client sends a "Hello_req" message to the server; then it waits for an answer. On receipt of a "Hello_res" message, the client goes to its final state.

---

6 For more details on transition notation, see section 15.3.7, Protocol Transition, in [SSU 10].
7 A complete description of state machine notation can be found in Chapter 15, "State Machines", in [SSU 10].

We can define this behavior by using a state machine. This kind of diagram allows us to represent the inputs and outputs clearly and rapidly. In our example, we define a state called "Idle" in order to indicate that the client class is waiting for an answer from the server. Note that the name of the state in this example is arbitrary and has no specific meaning.

Figure 2.12 shows a state machine diagram for the behavior of the client class.

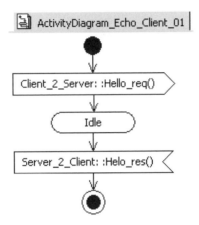

Figure 2.12. *Echo: state machine diagram for client class*

2.2.4.2. Server class

Now, let us model the behavior of the server class. The server starts in an "Idle" state and waits for a message from the client. On receipt of the "Hello_req" message, the server answers with a "Hello_res" message and then goes to its final state.

Figure 2.13 shows a state machine diagram for the behavior of the server class.

These two diagrams allow us to simulate and validate the echo protocol in the next section.

As you can see, this very simple example was quite easy to model by considering, in turn, the expected behavior, composite structure, communication interfaces and state machine diagrams.

In the rest of this chapter, we will extend this example in order to exchange the user data, first in a unidirectional way, and then in a full duplex way. We will also show how to model the expected behavior by using the behavior notation of state machine diagrams.

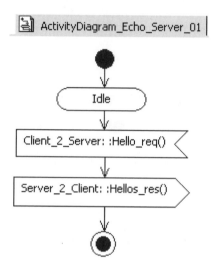

Figure 2.13. *Echo: state machine diagram for server class*

2.2.5. *Echo: validation and simulation*

Most of the time, the ultimate goal for an analysis and design team is to obtain a correct and complete model for a targeted system. Nevertheless, when working with big and complex systems, it is very common to forget some details. Moreover, people tend to neglect the exploration of all the possible alternate scenarios and all the error scenarios, because they often concentrate only on the complexity of the algorithms for the main scenario or the main business rules.

Simulation and validation allow the designer to act as a user, and to focus on a "test-drive" mode, which means concentrating test efforts on finding all the possible execution scenarios without taking into account the implementation details. It means conducting this work by considering it from a user's perspective.

We have used TAU G2 from Telelogic[8] in order to validate and simulate our models.

---

8 This software is now supported by IBM; however, when the simulations were performed, the software was supported by Telelogic.

2.2.5.1. *Simulating the model*

The bases for the simulation process are the activity and state diagrams. The simulation software automatically generates executable code from these diagrams; then, it executes the code and represents the simulation results as a sequence diagram.

The role of the designer during this phase concentrates on comparing the resulting simulation sequence diagrams with respect to those defined in the specification at the beginning of the analysis. If both diagrams match, it means that the model behaves as expected.

Figure 2.14 shows the sequence diagram resulting from the simulation. We can see that this diagram corresponds to that expected. We can therefore conclude that, under this very simple scenario, our model correctly corresponds to the expected behavior.

We should note the presence of an actor, "env". Our simulation software uses this actor to represent the "environment", which is a generic actor able to send messages to the objects in our simulation. We will see in later chapters how we can control our objects through this actor and how we can define more interactive scenarios.

We should also note that the messages sent between objects are asynchronous[9]. UML represents an "operation call" as a synchronous message and a simple "send" operation as an asynchronous message.

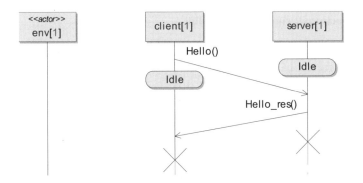

Figure 2.14. *Resulting sequence diagram
for the Echo protocol simulation*

---

9 A continuous-lined arrow with a filled in head represents a synchronous message. A continuous-lined arrow with an unfilled head represents an asynchronous message.

2.2.5.2. *Model validation*

The process just described allowed us to verify whether our model matches the specification or not. However, in larger models, there could be some parts of a model that do not behave as expected.

In order to validate model behavior, we need to consult the "coverage statistics" provided by some simulation tools. The analysis of these coverage statistics verifies that the executed scenarios fully explore the entire set of states and transitions of the state machine or of the activity diagram.

The coverage statistics for this model (Figure 2.15) show that our simulation visited 100% of the states and triggered 100% of the transitions in client and server classes. This fact shows that there are neither unvisited states nor unfired transitions in the state machines. Therefore, we know that there is no unexpected behavior in this case[10].

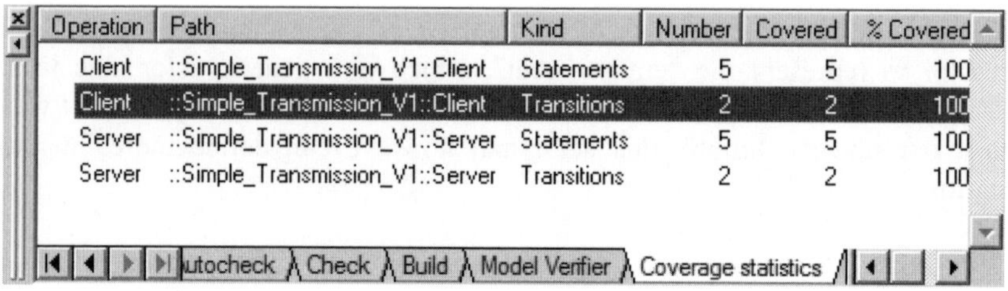

Operation	Path	Kind	Number	Covered	% Covered
Client	::Simple_Transmission_V1::Client	Statements	5	5	100
Client	::Simple_Transmission_V1::Client	Transitions	2	2	100
Server	::Simple_Transmission_V1::Server	Statements	5	5	100
Server	::Simple_Transmission_V1::Server	Transitions	2	2	100

utocheck ⋀ Check ⋀ Build ⋀ Model Verifier ⋀ Coverage statistics

Figure 2.15. *Coverage statistics for the Echo protocol*

2.3. Unidirectional: simple data sending

The preceding example showed how to model a very simple transmission protocol. On receipt of a message, the server only had to send a message back to the client and had then finished. However, it is possible to add many more functions to the server: it could have processed the message, extracted information, or stored statistics, for example.

In the same way, the first action of the client was to send a message to the server. Then, on receipt of an answer, it finished. However, it could have processed the received message, extracted information from it, or forwarded it to an application at a higher level.

---

10 This is true when no data values are used in the definition of the protocol behavior (in the behavior state machine). By contrast, when data is involved in the protocol behavior, for a full verification, we need to cover all the values of the concerned data, which is impossible in large systems.

2.3.1. *Requirement specification*

This second example aims to extend the behavior of the Echo example. This new protocol has the following characteristics:

– there are two communicating entities: client and server;

– client sends a Hello_req message to server;

– the Hello_req message contains a parameter of type Integer;

– the first message sent from client to server contains a parameter, "0", the second contains a parameter, "1", the third contains a parameter, "2", and so on;

– when receiving a Hello_req message, the server answers with a Hello_res message containing an integer;

– after sending the first Hello_req message, the client waits for a Hello_res message before sending a subsequent Hello_req message;

– the sending process continues indefinitely.

In fact, these requirements are quite ambiguous since they do not specify the relation between the data received and the data sent from the server side. For example, Figure 2.16 shows a communication diagram representing a possible sequence of messages fulfilling the previous specification.

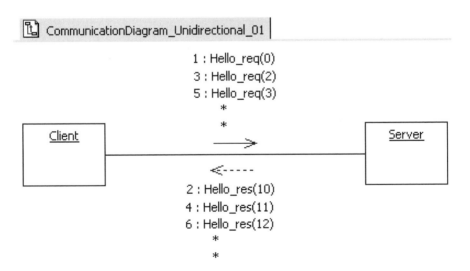

Figure 2.16. *Communication diagram for a unidirectional example*

However, this is not the correct sequence of the messages we expect to receive. Let us therefore change the specification and say that the server sends the integer data received back to the client. Let us also say that the client will increment the received data by one and will send the new value to the server again. These new requirements are specified as follows:

– there are two communicating entities: client and server;

– client sends a Hello_req message to server;

– Hello_req message contains a parameter of type Integer;

– the first message sent from client to server contains a parameter, 0;

– after sending the first Hello_req message, the client waits for a Hello_res message before sending the following Hello_req message;

– on receipt of a message, the client sends another Hello_req with the received parameter incremented by one;

– when receiving a Hello_req message, the server answers with a Hello_res message containing the parameter just received;

– the sending process continues indefinitely.

This new communication schema is represented in Figure 2.17.

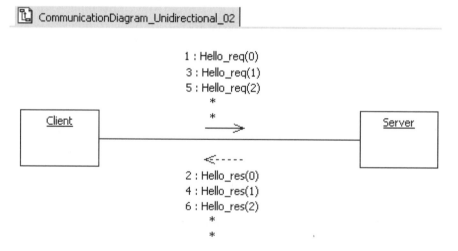

Figure 2.17. *Second communication diagram*
for a unidirectional example

As you can see, the most important value is the initial one since all the other values are calculated in runtime.

We could have defined the initial value as being given randomly or by a hypothetical user through a "user" interface. However, for simplicity's sake, we have not selected these options.

2.3.2. *Analysis*

2.3.2.1. *Sequence diagram*

In the same way as in the previous section, we can create a simple sequence diagram representing the expected behavior in a chronological order.

In Figure 2.18, the client sends a Hello_req(0) message to the server. On receipt, the server answers with a Hello_res(0) message. When the client receives the Hello_res message, it then takes the received message, increments the value by one and sends it back to the server.

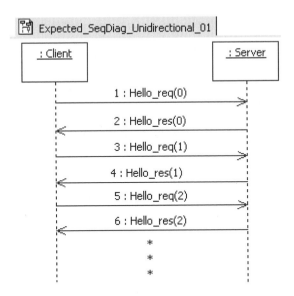

Figure 2.18. *Sequence diagram for a unidirectional example*

As you can imagine, the initial class diagram corresponding to this new protocol is just the same as the Echo one (see Figure 2.3).

2.3.2.2. *Signals list definition*

We can use the same strategy as that used for the Echo protocol, gathering the messages going from one node to another into interfaces. In order to keep some similarity with the previous example, we can call the interfaces Server_2_Client and Client_2_Server. Moreover, they will contain the same messages. The only difference will be that the messages will contain a parameter. Figure 2.19 shows these interfaces.

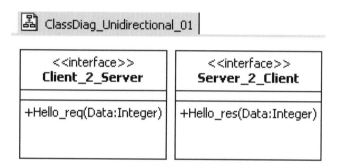

Figure 2.19. *Interfaces for a unidirectional protocol*

2.3.3. *Architecture design*

Now, we can add a port P_C to the client and a port P_S to the server, and then link the interfaces to the ports. Finally, the ports are connected through a communication channel CH_C-S. The resulting diagram is shown in Figure 2.20.

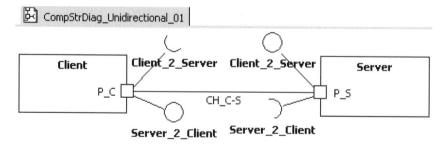

Figure 2.20. *Client and server with ports, interfaces and channels*

2.3.4. *Detailed design*

2.3.4.1. *First variation: protocol notation (transition oriented)*

It is easy to see that the state machine diagram representing the server's behavior is almost the same as that in the previous example. The main difference is that the server has to receive a parameter and then to send it back to the client.

Note that the server does not need to really process the parameter received. We can change the server's diagram in the echo example and add the requested parameter. Figure 2.21 shows the new behavior for the server.

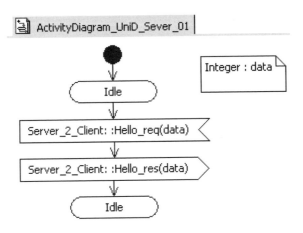

Figure 2.21. *State machine for server class (transition oriented)*

Note that, for the client class, there is a transition triggered without any signal. The initial transition is usually the only transition allowed to be triggered without any kind of stimuli or guard condition. This initial transition sends the very first Hello_req message to the server, and then defines the beginning of the sequence to be used; in our case, this first value is set to 0. Figure 2.22 shows the client's behavior.

It should be noted here that the "idle" state is referred to twice in the same diagram. In fact, there is a single state being referred to twice. The reference at the bottom means that the system goes back to the idle state after sending the incremented data to the server. This representation allows us to avoid drawing long and intricate transitions.

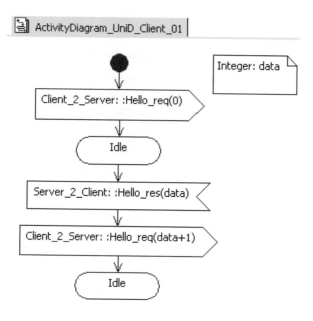

Figure 2.22. *New state chart diagram for client class*

2.3.4.2. *Second variation: behavior notation (state oriented)*

The diagrams used to define the behavior of the client and server classes in the preceding section use a transition-oriented design; however, we could have used a different UML style: state-oriented design.

This time, we can use a transition with a guard condition to send the first message (see note (1) in Figure 2.23). This transition is not triggered by an external stimulus, but by an internal condition. Indeed, every time the client is in an idle state, if the condition is true, then the transition will be triggered. Note that, in order to avoid a repeated sending of the Hello_req(0) message, the client increments data after the first message, and the notation for sending a signal is ^signal-name(parameters).

A second transition accepts the messages coming from the server and sends a message back to the server with incremented data (2). Figure 2.23 shows this behavior.

Note that we define a parameter named "data" by using a text-note. We also initialize the parameter in the same text-note. Also note that every message refers to the interface in which it was declared. This reference is not mandatory; however, it might be very useful in a big model, where many variants of the same message might be declared in different interfaces. This referencing is done by using traditional object-oriented syntax: ClassName::referredElement.

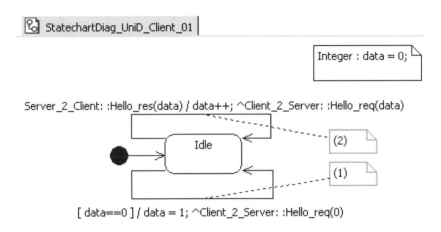

Figure 2.23. *State machine for a unidirectional client*

Finally, in this diagram, note the use of the semicolon (;) indicating a sequence of actions, conditions and/or outputs: e.g. action1; action2; output1; output2; etc.

We can refine this diagram by removing the transition with the guard condition and using the initial transition in order to send the initial message (see (1) in Figure 2.24).

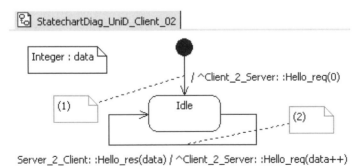

Figure 2.24. *State machine for a unidirectional client (2)*

We can also think about removing the effect on the second transition and incrementing the received data directly in the output (see (2) in Figure 2.24). This final state machine is shorter than the one shown in Figure 2.22; however, it is harder to understand.

The state machine diagram representing the server's behavior is quite easy to build. It contains a simple transition that, on receipt of the Hello_req message, sends the received data back to the client; see Figure 2.25.

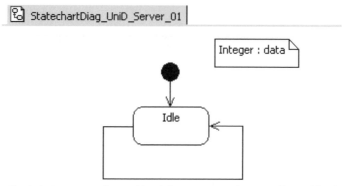

Figure 2.25. *State machine for a unidirectional server*

As we can see, the transition-oriented design is easier to understand than the state-oriented design since the inputs, outputs and actions are explicitly and graphically represented; however, it is longer. Nevertheless, these two representations are functionally equivalent.

For simplicity's sake, we will generally use the transition-oriented design from this point onwards.

2.3.5. *Validation and simulation*

2.3.5.1. *Simulating the model*

We begin by simulating the transition model given in section 2.3.4.1 and comparing it with the sequence diagram described in Figure 2.18.

The resulting sequence diagram is shown in Figure 2.26.

Note that the sequence of messages obtained in both diagrams is the same. The main difference is that, in the simulation, we see the state in which the class is left after an action is executed. Of course, this information was not available when the first sequence diagram was created and it can only be obtained after a detailed design.

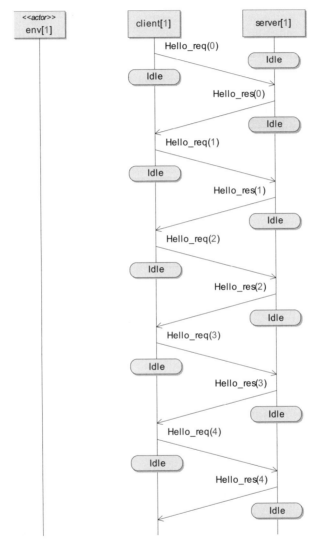

Figure 2.26. *Resulting sequence diagram*
for a unidirectional protocol simulation

2.3.5.2. *Model validation*

At this point, we know that our model behaves as expected under this scenario.

By looking at the coverage statistics, we can be certain that our scenario correctly allows us to visit all the states and to trigger all the transitions in both classes.

Figure 2.27 shows that all the transitions and all the states in both classes in our model have been covered by our test scenario.

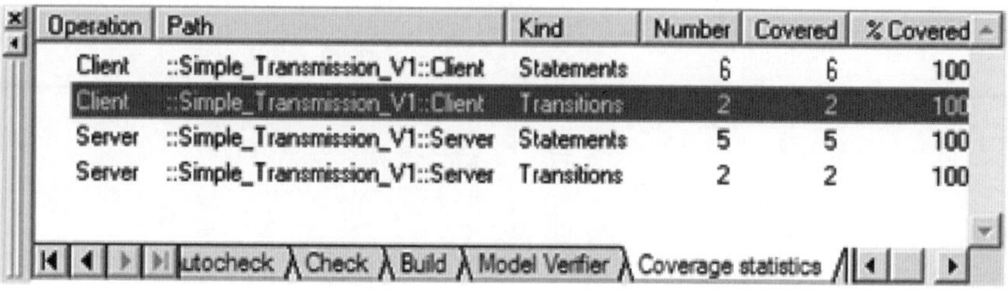

Operation	Path	Kind	Number	Covered	% Covered
Client	::Simple_Transmission_V1::Client	Statements	6	6	100
Client	::Simple_Transmission_V1::Client	Transitions	2	2	100
Server	::Simple_Transmission_V1::Server	Statements	5	5	100
Server	::Simple_Transmission_V1::Server	Transitions	2	2	100

Autocheck ∖ Check ∖ Build ∖ Model Verifier ∖ Coverage statistics

Figure 2.27. *Coverage statistics for a unidirectional protocol*

2.4. Full duplex: simple data sending

One important advantage of using UML is its ability to reuse an existing model in order to create a more complex new model. In this section, with a new example, we will create a bidirectional communication. Note that the first step of this exercise does not take into account any possible error in the medium, such as delay, losses, misorder, etc.

2.4.1. *Specification*

This example aims to extend the behavior of the "unidirectional – simple data sending" protocol defined in the previous section. This new protocol has the following characteristics:

– there are two communicating entities: User1 and User2;

– the behavior of both users is identical;

– UserX sends a "Hello_req" message to UserY;

– Hello_req message contains a parameter of type Integer;

– the first message sent from UserX to UserY contains a parameter, 0;

– when receiving a Hello_req message, UserY answers with a "Hello_res" message containing the parameter just received;

– on receipt of a Hello_res message, UserX sends back another Hello_req with the received parameter incremented by one;

– after sending the first Hello_req message, UserX waits for a Hello_res message before sending the following Hello_req message;

– when UserX is waiting for a Hello_res message, it is able to answer a Hello_req message;

– the sending process continues indefinitely.

Many different combinations are possible; for example, only User1 or User2 send data. Figure 2.28 shows these two combinations.

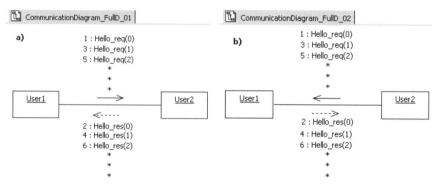

Figure 2.28. *Communication diagram for the full duplex protocol*

It is also possible to imagine that both users send data simultaneously. In fact, the specification neither indicates which user starts sending data, or the sending frequency. It is possible to imagine that the sending rate of either of the users is different from that of the other.

2.4.2. *Analysis*

We can represent the expected behavior through a sequence diagram and can use it to imagine a more complex scenario where the users do not start at the same time and where they transmit at different rates.

2.4.2.1. *Sequence diagram*

Figure 2.29 represents a possible scenario for a full duplex protocol, where both users transmit data in a full duplex mode.

In this diagram we can see that messages 4, 6, 10 and 12 correspond to the data sent by User2 to User1; the other messages correspond to data sent by User1 to User2.

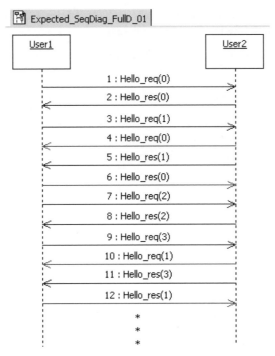

Figure 2.29. *Expected behavior for a full duplex protocol*

We can also note that User1 sends data at a higher rate than User2 and that User2 started sending data later than User1 (its first message is the fourth message in the sequence).

As you can see, the messages exchanged between the users correspond to those used for the unidirectional protocol. Thus, we can reuse the interfaces defined in Figure 2.19 in section 2.3.2.2.

Let us first analyze the architecture representing this new behavior; after that, we will link the interfaces to the ports concerned.

2.4.2.2. *Concerned classes*

In the previous exercises, the client sent a Hello_req message and the server answered with a Hello_res message. In this current protocol, the communication becomes bidirectional. Both users now exhibit the behavior of both client and server. In this case, client and server classes are now not the main communicating elements, but components of the user class. We will now see how we can model this behavior.

We could have considered an inheritance relationship, where a user specializes[11] client and server classes. However, a multiple inheritance is difficult to manage with most of the modern programming languages. Moreover, it would have been necessary to manage a merge of both state machines.

Instead, we prefer to chose a different solution: defining a composition relationship between user, client and server. In this way, each object can run in a separate thread. Figure 2.30 shows this composition relationship.

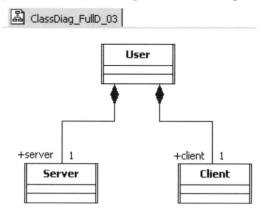

Figure 2.30. *A user behaves as a server and a client at the same time*

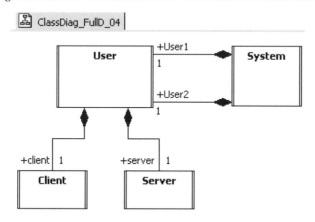

Figure 2.31. *Defining two instances of user class*

---

11 Specialization means creating new more specific subclasses from an existing class. This is a different point of view from "generalization" which means extracting shared properties from a set of classes and gathering these properties into a super class. Generalization/specialization are also usually known as "inheritance". For more details on this association, see 7.3.20, Generalization (from Kernel, PowerTypes) in [SSU 10].

We can complete this diagram by creating two different instances of the user class, instantiated by a system class as shown in Figure 2.31.

2.4.3. *Architecture design*

We have already defined a port for the client and server classes and have linked those ports to corresponding interfaces (see Figure 2.32).

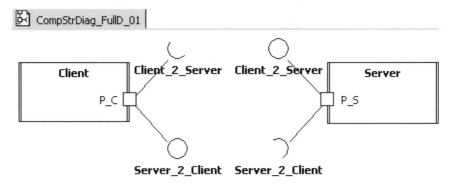

Figure 2.32. *Client and server classes with ports and interfaces*

We have defined a user class, composed of a client and a server. We can view this composition as if the user class encapsulates and isolates the client and server classes. This means that these components will not have any direct interaction with any element outside the user. This composed architecture is represented in Figure 2.33.

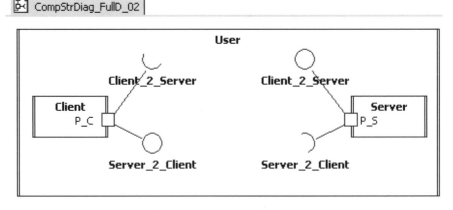

Figure 2.33. *Client and server classes isolated inside user class*

Notice that the client and the server are isolated from the external world; thus, in this context, they cannot receive or send any message to any entity outside the user. However, if we add a Service Port[12] P_U to the user, we can create a communication point enabling the user's components to send and receive messages to/from entities outside the user. After that, we can link the new port to the internal parts by using two communication channels: CH_C-U and CH_S-U. These channels carry messages passing through P_U port. Finally, we can link the interfaces to this new port in order to keep communication coherence. These modifications are represented in Figure 2.34.

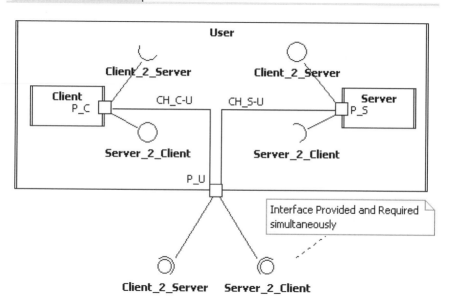

Figure 2.34. *User class with port and interfaces*

Notice that this new port realizes and depends on Client_2_Server and Server_2_Client interfaces at the same time. This means that P_U port uses each interface as input (realization) and output (dependency). The symbol linked to the text-note indicates that the interface is provided (realized) and required (dependent) at the same time.

---

12 There are two kinds of ports in UML: service ports and behavior ports. Service ports only relay received messages to an internal part. In a behavior port, the received messages are sent to the behavior (activity, state machine, etc.) of the class owning the port.

Now that we know the internal architecture of the user class, we can describe more precisely the sequence diagram presented in Figure 2.29 by taking into account the parts composing each user (see Figure 2.35).

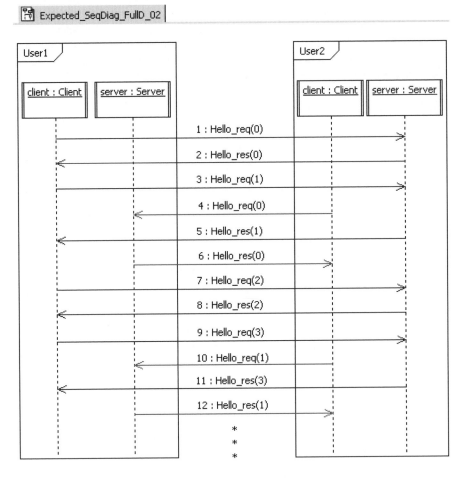

Figure 2.35. *Expected behavior for the Full Duplex protocol,*
users classes and their parts

It is easy to see that both data flows are concurrent and that one client does not know the client on the other side. The same assertion is valid for the servers. For clarity's sake, we can represent each flow in a separate diagram, as shown in Figure 2.36.

As you can see, we have reused the already modeled client and server classes and state machines. The user class has no particular behavior; it only represents the fact that two instances of user class are instantiated at the beginning of the system.

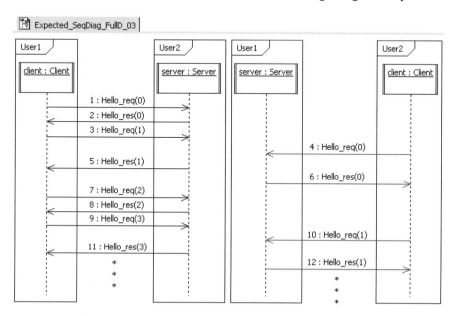

Figure 2.36. *Separated and parallel data flows
between User1 and User2*

2.4.4. *Validation and simulation*

2.4.4.1. *Simulating the model*

Figure 2.37 shows the beginning of the sequence diagram obtained from the simulation of our model. Notice that both clients start sending a Hello_req message; then, all the components go to their idle state as expected.

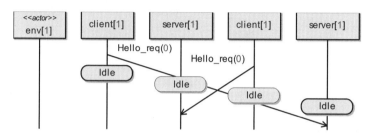

Figure 2.37. *Resulting sequence diagram for a full duplex protocol*

Figure 2.38 shows a longer execution.

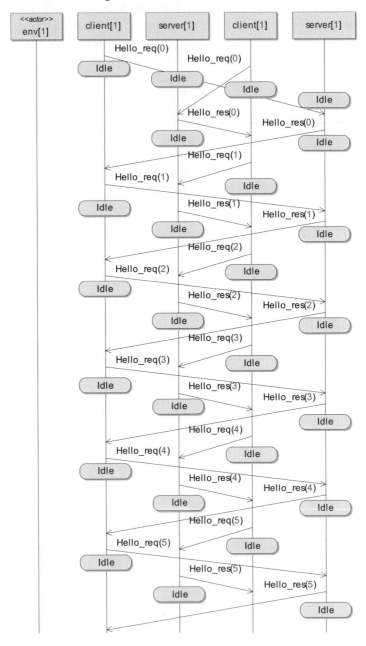

Figure 2.38. *Resulting sequence diagram*
for a full duplex protocol simulation

In Figure 2.38 we can see the elements composing User_1 and User_2. This sequence diagram correctly matches the expected behavior defined in section 2.4.2.1; however, User_1 and User_2 are not easy to compare. Let us repeat the simulation using a different granularity in order to hide the internal structure of user classes. Figure 2.39 shows this new simulation.

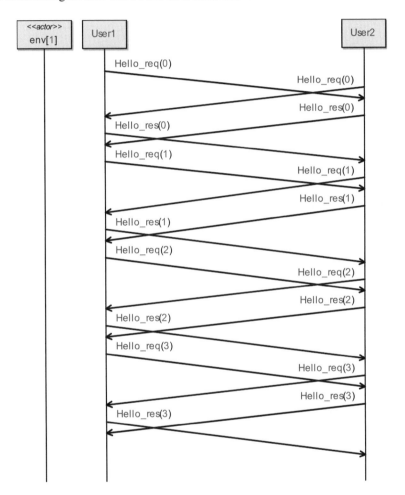

Figure 2.39. *Resulting sequence diagram*
for a full duplex protocol simulation (simplified)

This sequence diagram is much more similar to the one defined in the analysis phase than Figure 2.38 was. We can now confirm that the behavior obtained corresponds to that specified.

2.4.4.2. *Model validation*

If we look at the coverage statistics, we can see that, even if there are two instances of the client class and two instances of the server class running in our simulation, the state machine is the same for each instance of the same class; thus, there are only eight statements and two transitions for the client class and five statements and two transitions for the server class.

We can also see, in Figure 2.40, that our simulation scenario has visited all the states and triggered all the transitions. There are, therefore, no unexpected behaviors.

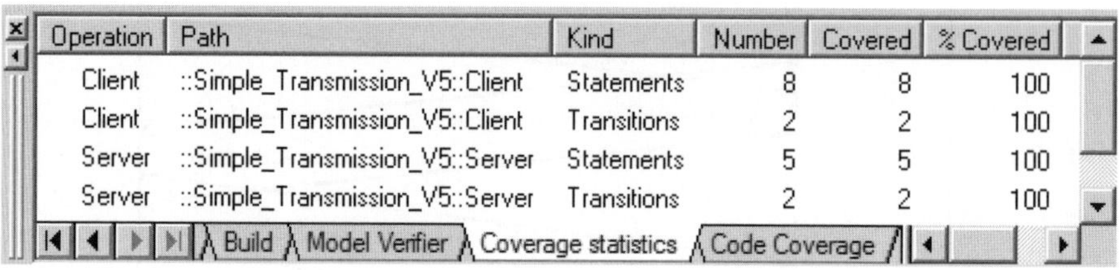

Operation	Path	Kind	Number	Covered	% Covered
Client	::Simple_Transmission_V5::Client	Statements	8	8	100
Client	::Simple_Transmission_V5::Client	Transitions	2	2	100
Server	::Simple_Transmission_V5::Server	Statements	5	5	100
Server	::Simple_Transmission_V5::Server	Transitions	2	2	100

Build ⅄ Model Verifier ⅄ Coverage statistics ⅄ Code Coverage

Figure 2.40. *Coverage statistic for the full duplex protocol*

2.4.5. *Different ways of doing the same thing*

In section 2.4.2.2 we proposed composing a user class based on the existing client and server classes. However, this is not the only solution. In fact, many other solutions are possible, each one with its benefits and its drawbacks, and we will now consider some of them.

2.4.5.1. *A single machine*

The first solution could consist of creating a single state machine for the entire user class. This single machine would have to manage sending and answering processes at the same time.

We can begin with the state machine diagram shown in Figure 2.41 (functionally the same as the one describing the client's behavior for the unidirectional protocol described in Figure 2.22 in section 2.3.4.1). This diagram describes the behavior needed in order to manage the sending process, i.e. the client.

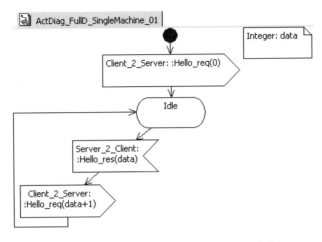

Figure 2.41. *Full duplex protocol: client's main behavior*

The solution entails extending this client state machine in order to manage the answering process, i.e. the server.

We can reuse the Idle state and the data integer parameter in order to answer any request. Figure 2.42 shows a single extended state machine managing client (left) and server (right) behavior at the same time.

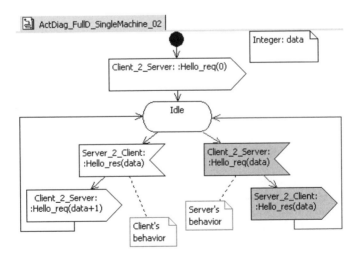

Figure 2.42. *Full duplex protocol: single state machine performing both client's and server's behavior*

Note that this single state machine is able to fulfill our requirements perfectly and would not be so hard to implement. However, many drawbacks can be detected.

First, we have already proposed a complete client and server class in section 2.3, and have seen that creating a composite of both of them is very simple. Furthermore, we consider that it is better to reuse the existing model instead of creating a new one.

In our current example, the base behavior (client behavior) is quite simple: it receives a response message and sends a new request back to the server with a new parameter; this receive-send behavior is performed on a single transition. Now, imagine that the client's behavior is much more complex, i.e. that it contains several states and triggers complex actions. In that case, extending such a state machine would be a very hard task.

In order to illustrate this situation, let us suppose that the imaginary state machine shown in Figure 2.43 represents the client's behavior. This state machine is a little more complex than the original one (shown in Figure 2.41) since the new state machine contains several states and transitions.

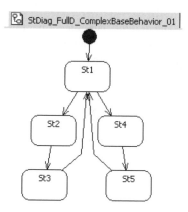

Figure 2.43. *Full duplex, with more complex client behavior*

Under these circumstances, the extension needed to implement the server's behavior in such a state machine is not as simple as in the previous example. Indeed, a very strong constraint in our system is that it should be ready to answer a request message at any moment; otherwise, the received message might be lost.

In order to combine the server's behavior in this imaginary client state machine, we need to repeat the extension made in Figure 2.42 for every single state in the client state machine. Figure 2.44 shows the state machine resulting from the combination of the new imaginary state machine and the server's behavior. Note that the transition representing the (simple) server's behavior was copied to every single state of the state machine representing the client's behavior.

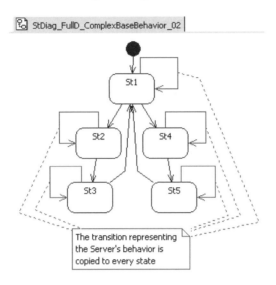

Figure 2.44. *Full duplex: extending complex client behavior*

Imagine now that the server's behavior needs to evolve. Now, when the server receives a request, it runs some internal complex behavior, and then answers the request. Thus, we need to update our state machine in order to take this new behavior into account. If we follow the same process as the one described in the preceding figure, we will also include the new server's complex behavior in every single state in the client's behavior. Figure 2.45 shows a possible resulting state machine. In this diagram, the states labeled St1 to St5 (in the center) represent the client's behavior, while the states labeled StE1 and StE2 represent the server's behavior.

It can be seen that the resulting machine has become much more complex than what we imagined in the beginning. Moreover, every single evolution in either the client or the server will heavily affect the model (and the implementation). Indeed, if we add a new state to the client, we need to copy the entire server's behavior to the new state. In the same way, if we modify the server's behavior, then we need to update every copy of the server's state machine.

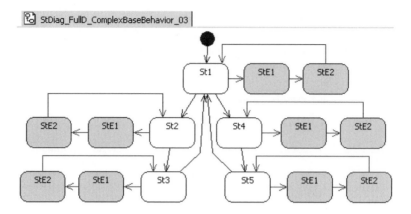

Figure 2.45. *Full duplex: complex client and complex server in a single state machine*

As a consequence, any evolution leads to a possible strong modification of the state machine. A good solution to this problem is to completely separate both behaviors in order to modify them independently from one another. This solution reduces the impact of any modification and confines it to a specific independent region, and we will now look at how we can define this kind of solution in UML.

2.4.5.2. *Interruption*

A possible solution to separate the client and server behaviors is to use an interruption.

UML proposes two representations for modeling interruptions: interruptible regions and composite states. An interruptible region is an artifact defined in activity diagrams; it represents a "region" containing a set of activities which can be interrupted at any moment by a given input. When the region is interrupted, all the control flows contained by the region are aborted[13]. Figure 2.46 shows an example extracted from the current UML specification [SSU 10].

In our case, we need to keep track of the state of the system when the interruption is received in order to come back when the interruption is finished. This constraint leads us to investigate the second solution, using composite states.

---

13 More details on interruptible regions can be found in section 12.3.33, Interruptible Activity Region, in [SSU 10].

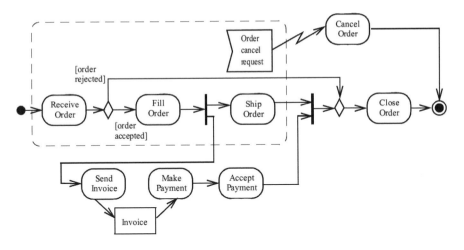

Figure 2.46. *Example of an interruptible region*

A composite state is a state composed of a sub-state machine. From a high level view, a composite state behaves just like any other state. From a low level view, a composite state merely defines a complex behavior.

In Figure 2.47 you can see an example of a composite state extracted from the current UML specification [SSU 10]. In this diagram, the composite state is called "Active". This state is composed of a sub machine. When this state becomes active, the internal sub-state machine is executed. At any moment during the internal sub-state machine execution, if the system receives an "abort" signal, the control flow goes to the "aborted" state (see note (1) at the bottom of the diagram).

UML also proposes a very effective means of storing the last active state before the interruption: a "shallow history" pseudo-state. A shallow history pseudo-state is used in UML to represent the most recent active sub-state of its containing state and it is represented as a circle with the letter "H" inside [14].

We will now define a composite state and call it MainBehavior. This state is composed by a sub-state machine implementing the client's complex behavior. Figure 2.48 shows the MainBehavior composite state with the contained state machine.

---

14 For more details on the shallow history pseudo-state, see section 15.3.8, Pseudostate, in [SSU 10].

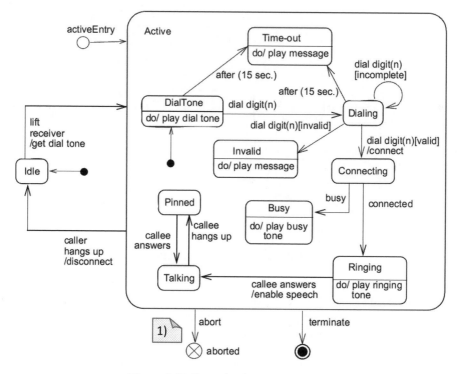

Figure 2.47. *Example of a composite state*

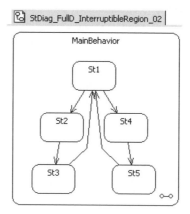

Figure 2.48. *Full duplex: composite state for client's behavior*

We need to extend the main behavior with the server's behavior. From the point of view of the client, the server's behavior will be perceived as an interruption. After the interruption is finished, the system will come back to the previous internal state within MainBehavior. We can model this behavior by using the Shallow History pseudo-state, as shown in Figure 2.49.

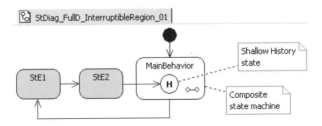

Figure 2.49. *Full duplex: extending*
a composite state with an interruption

This new state machine allows us to separate the client's behavior from the server's behavior, and Figure 2.50 represents both behaviors in a single diagram. Note that, if the main behavior is interrupted in St4 for example, after the interruption, the system comes back to the same state thanks to the utilization of shallow history pseudo-state.

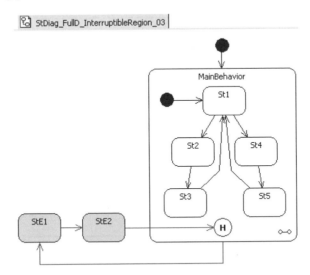

Figure 2.50. *Full duplex: composite state*
with internal sub-machine and interruption

Imagine now that the server behavior is much more complex than that represented in Figure 2.50. In this case, the interruption might last for a long time;

then, the Shallow History pseudo-state representing the most recent active sub-state of its containing state, might become outdated. At that time, the messages sent by the main behavior might be out of date or useless. This is particularly true in real-time system, such as telecommunication protocols.

The solution to this problem can be found by executing both behaviors, not as an interruption (which is to say, sequentially) but as parallel behaviors.

2.4.5.3. Composite states with concurrent regions

A good solution to represent parallel behaviors in UML is the use of a composite state machine with concurrent regions.

State machines with concurrent regions make it possible to express complex behaviors where two or more states can be active simultaneously. In such a case, many states (one from each region) might be active at the same time.

A state with concurrent regions is represented by dividing the state by a dashed line. Each region may optionally have a name and contain an entire state machine. Figure 2.51 shows an example of a composite state with concurrent regions extracted from the current UML specification[15] [SSU 10].

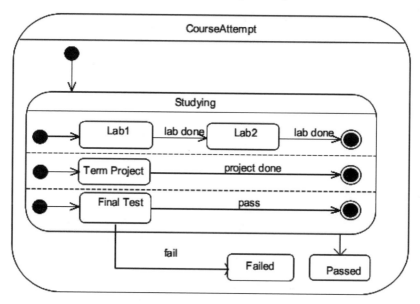

Figure 2.51. *Example of a composite state with concurrent regions*

15 For more details on composite states with concurrent regions, see section 15.3.11, State, in [SSU 10].

In our example, all we need is to separate client and server behaviors and place them into separate regions. With this solution, both state machines representing a different behavior will be executed simultaneously. In this way, the user can answer any request, complex or not, while waiting for an answer to its own request.

As you can see, this solution allows us to modify any of the state machines without impacting the others. It also allows us to avoid long interruptions and thus to avoid possible outdated processes. Figure 2.52 shows a composite state with concurrent regions.

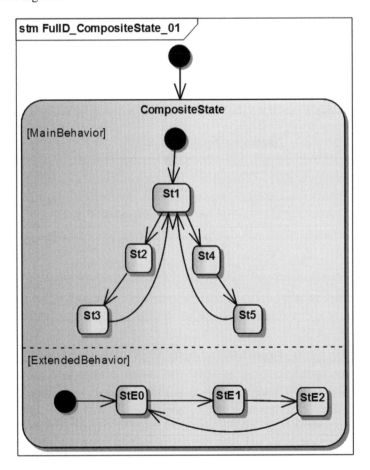

Figure 2.52. *Full duplex: composite states with concurrent regions*

Note that each internal state machine contains its own initial state.

2.4.5.4. *Parallel threads*

Using a composite state with concurrent regions is not the only way to represent parallelism in UML.

Another way to represent parallelism is by using forks. A fork is a UML artifact very similar to the formalism used in Petri Nets, and it is represented as a short thick line with an incoming edge and two or more outgoing edges. The edges (symbolized by an open arrow head line) coming into and out of a fork node represent either object flows or control flows. An incoming control token is duplicated in each of the outgoing states (i.e. places in a Petri Net). The fork (and its counterpart: a join) are used in order to represent the split of a flow into multiple concurrent flows[16] (and to merge concurrent flows).

Figure 2.53 shows our model split into parallel state machines by using a fork. Note that we again find both behaviors separated one from the other. This separation allows us to update one of the state machines without impacting the other one.

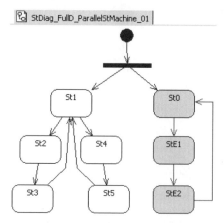

Figure 2.53. *Full duplex: parallel threads*

This seems to be a good solution to our bidirectional communication system. However, there is a drawback to it: implementing parallelism implies using multithreading which can involve complex code. Moreover, both behaviors are placed into the same class and, thus, the risk of side-effects when modifying one of the state machines is incremented.

We still prefer our first solution, i.e. thinking about using many instances of a single class (client and server) instead of handling a single class. In other words, we

16 For more details on fork and join nodes, see section 12.3.30, ForkNode, in [SSU 10].

prefer to leave the responsibility of managing any parallelism required to the operating system rather than to the programmer.

2.5. Chapter summary

This chapter presented a simple transmission protocol concerning a client and a server. This case study allowed us to show that it is possible to think about protocols through an object-oriented model, and particularly by using UML 2.

We then showed how reutilization is easy to represent, describe and use by defining a well-structured model.

In the following chapters, more features will be added to the models by including an application and a protocol, as well as a transmission medium.

2.6. Bibliography

[DOL 03] DOLDI L., *Validation of Communications Systems with SDL: The Art of SDL Simulation and Reachability Analysis*, John Wiley & Sons Ltd, New York, USA, 2003.

[OCL 10] Object Constraint Language, Version 2.2, OMG specification, 2010-02-01, http://www.omg.org/spec/OCL/2.2

[SSU 10] Unified Modeling Language™ (UML®), Superstructure specification, 2010-05-05, http://www.omg.org/spec/UML/2.3/

Chapter 3

Simple Chat Application

3.1. Introduction

This chapter presents a simple chat application.

We will connect this chat application with a reliable transport protocol in a further chapter. The main goal here, however, is to familiarize the reader with the proposed roadmap to be used for the analysis and design of applications and protocols through a simple chat example. The chapter also presents the life-cycle schema to be used in the rest of this book.

We think that this is a representative example of a simple but useful protocol application. In the next chapters, we will model two layers, representing the transport protocol and a basic transmission medium.

3.2. Requirements

Two remote users (User_A and User_B) want to communicate by using a simple chat application. Each user has an instance of the application:

– A user can:

- open a session;

- accept a session;

- refuse a session;

- close a session;

- send data (in order to simplify the example, the data to be sent is a Char String).

- A session can be opened by either User_A or User_B.

- When a user asks the system to open a session, the system sends a message to the other user in order to ask him/her to either accept or refuse the session.

- When a user is asked to accept a session, if no answer is received after a given time, the system will cancel the request and inform the requesting user that no answer was received.

- The system will inform the requesting user of the answer provided by the requested user (either accept or refuse).

- The system will inform the users whenever the chat partner closes the session.

- When a user sends data, the system will inform that user about the sending status:

- data sending OK; or

- data sending Error.

- The system will transfer data only when a session is opened.

3.3. Analysis

We can represent the given requirements as a scenario modeled with an activity diagram (see Figure 3.1).

User_A opens a session (1). Then, User2 can either accept or refuse the session.

After a given time (TimeOut event), the system cancels the request if no answer is received; then, the control flow goes to the final state (2). Otherwise, if User_B refuses the session, the control flow goes to the final state.

If User_B accepts the session, the control flow allows both users to either send data (3) or close the session. Note that thanks to the fork node, both users can send data at the same time. Nevertheless, a side effect of this is that any of the users can close the session while the other is still sending data. This might produce an error that we will need to manage.

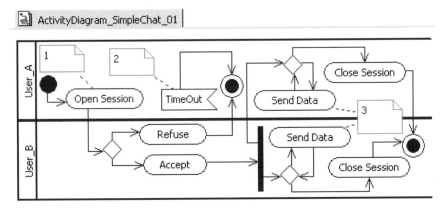

Figure 3.1. *Simple chat. State machine diagram for Simple Chat, with only user activities*

Figure 3.1 correctly fulfills the given requirements; however, it only represents the actions performed by human users; it does not represent the actions of the system, i.e. mainly notifications of requests sent by users, timer reset, and request cancellation after timeout. The necessity of representing the system is even more evident when we think about the timeout. This event represents the receipt of an interruption signal coming from a clock and it is, of course, not received by the users themselves, but by the system.

We can therefore improve on Figure 3.1 by adding system activities to it (see Figure 3.2).

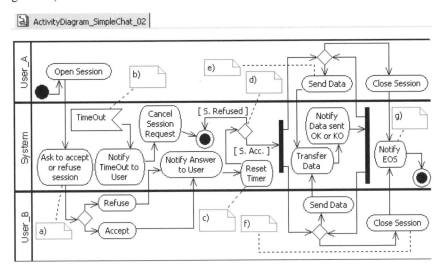

Figure 3.2. *Simple chat. Activity diagram for simple chat, users and system activities*

Note that when we refer to the "system", we refer to a black box representing all the entities linking the users, i.e. the applications, protocols and the network.

Note that, in Figure 3.2, after the "Open session" activity is performed by User_A, the system asks User_B to either accept or refuse the session (note (a)). If no answer is received after a given time (note (b)), the system notifies User_A about the TimeOut and cancels the request. Otherwise, the system informs User_A about the answer given by User_B and then resets the internal timer (note (c)).

After that, the system goes to the final state if User_B refuses the session. If the session request is accepted (note (d)), then the system allows both users to either send data (note (e)) or close the session (note (f)). When a user sends data, the system transfers the data and notifies the sender about the result (OK or KO). When a user closes the session, the system notifies the other user (note (g)) of the end of session (EOS), and then goes to its final state (represented here by the Flow Final[1] node).

If we compare Figures 3.1 and 3.2, we can see that the activities performed by the users are the same: open session, refuse, accept, send data and close session. We will base our model on these scenarios and will also use them in order to test it.

3.3.1. *Sequence diagrams*

In order to create the state machines representing the behavior of the chat application, it would be very helpful to have the list of messages exchanged between users and the system; additionally, we need to have this list in a chronological order and organized by scenario.

The easier method to obtain this list of messages is to create a sequence diagram based on the activity diagram in Figure 3.2. The problem is that a single sequence diagram representing all the activities and all the possible combinations might be very long and difficult to analyze.

In order to avoid a huge sequence diagram, we will use an InteractionUse[2] for each of the user actions described in Figure 3.2. Then, each InteractionUse will be

---

1 For more details on control nodes, see section 12.3.20, Control Node (from Basic Activities), in [SSU 10].

2 The InteractionUse is shown as a CombinedFragment symbol where the interaction operator is called "ref". An InteractionUse refers to other interactions, for example, other sequence diagrams. This is particularly useful for splitting a huge diagram or sharing a specific interaction fragment between multiple diagrams (factorization). A CombinedFragment is an expression or specialization of an InteractionFragment, and is always defined by its interaction operator (ref, alto, opt, par, break, else, seq, strict, neg, etc.). An InteractionFragment is a piece of an interaction.

described separately, using more sequence diagrams. Finally, we will be able to gather all the messages and classify them into Interfaces.

Figure 3.3 shows that the scenario starts when a user opens a session. Then, three exclusive situations may occur: refuse, timeout or session establishment. Note that the exclusive execution of these scenarios is guaranteed by the "alt" interaction operator.

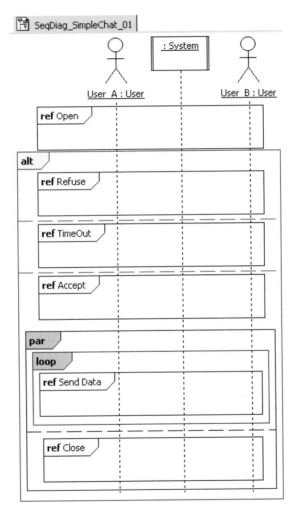

Figure 3.3. *Simple chat. Main expected sequence diagram*

After session acceptance, the users send data and then any of the users close the session. Note that "send data" and "close" might happen simultaneously, i.e. one user might be sending data while the other is closing the session. The parallel execution of these scenarios is guaranteed by the "par" interaction operator. The ability of both users to execute an action while the other is executing a different action might lead to an error. This kind of error will be discussed and managed later in this chapter.

Having organized the possible scenarios chronologically, we are now ready to discover the messages exchanged inside each InteractionUse.

3.3.1.1. *Splitting the system up*

First of all, it is useful to look inside the system. As defined in section 3.2, each user utilizes a chat application. We know that both applications are an instance of the same class. From this, we can guess that, from a high level point of view, the system is composed of two instances of the same class, which we will name "ChatApplication".

We can then split the system into two separate objects: Chat_App_A and Chat_App_B. These objects are instances of class ChatApplication and represent the chat application used by each user. Figure 3.4 shows these two new objects.

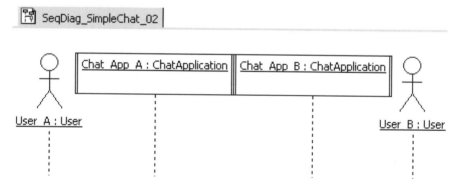

Figure 3.4. *Simple chat. Splitting the system class*

Note that this example does not represent the other elements composing the system: the protocols and the medium. These elements will be modeled in further chapters. For the moment, we will only consider a reliable data transmission.

3.3.1.2. *Open session*

Figure 3.5 represents only the opening session request and not the possible endpoints (refuse, accept and timeout). These possible endpoints will be explained in further diagrams.

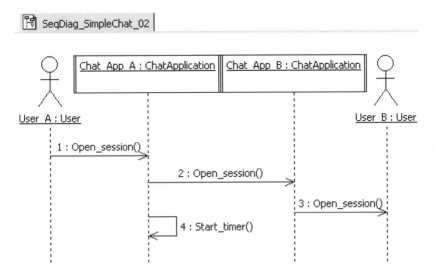

Figure 3.5. *Simple chat. Expected sequence diagram for "open" session*

The requirements described in section 3.2 indicate that any of the users can open a session. In Figure 3.5, User_A sends a request to his chat application (message 1). Then, the requirements indicate that the system has to ask the other user to accept, or not, the session (messages 2 and 3). After that, if no answer is received after a given time, the request is cancelled. The way to implement this behavior is to set a timer which will be read by the system (message 4).

3.3.1.3. *Timeout*

Figure 3.6 shows the expected behavior for a timeout. As we have said, after a given time, if no answer is received from the second user (message 1), then the requesting application must cancel the request (message 2) and inform its user (message 3).

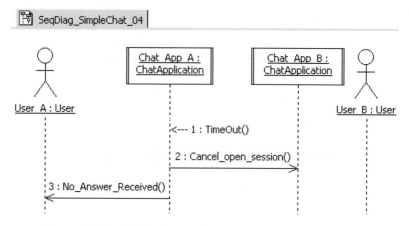

Figure 3.6. *Simple chat. Expected sequence diagram for "timeout"*

3.3.1.4. *Refuse*

Figure 3.7 shows the expected sequence diagram for the refuse session. As you can see, User_B refuses to accept the session request from User_A (message 1). Then, chat application 2 informs the requesting application about the answer (message 2) by sending an Open_session_ref message. This is only one solution among different possibilities. Another solution would have been to send a more generic message, for example Open_session_res(parameter_list) with a parameter indicating the nature of the answer (accept or refuse), and probably a few other parameters that are useful for correctly establishing the connection. However, for simplicity's sake, we have decided to use a different message and not to send any parameter within the answer message.

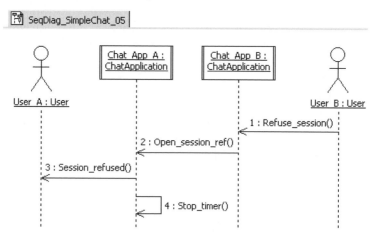

Figure 3.7. *Simple chat. Expected sequence diagram for "refuse" session*

On receipt of the Open_session_ref message, Chat_App_A informs its user about the answer (message 3) and, finally, resets the timer set in section 3.3.1.2 (message 4).

3.3.1.5. *Accept*

Figure 3.8 shows the expected sequence diagram for accepting a session. User_B accepts the session (message 1). Chat_App_B informs the requesting application about the answer by using Open_session_ok (message 2). After that, the requesting application informs its user about the answer (message 3). Finally, Chat_App_A resets the timer previously set in section 3.3.1.2 (message 4).

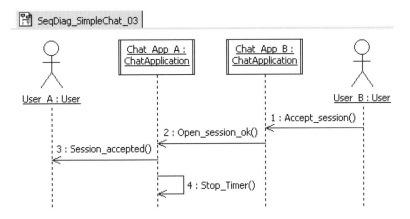

Figure 3.8. *Simple chat. Expected sequence diagram for "accept" session*

3.3.1.6. *Close*

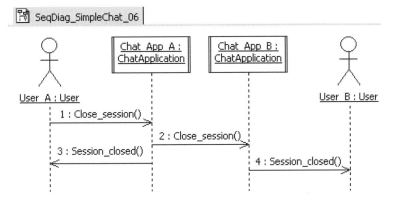

Figure 3.9. *Simple chat. Expected sequence diagram for "close" session*

Figure 3.9 shows the case where User_A closes the session (message 1). His application informs Chat_App_B (message 2) and then informs its own user that the session has been closed, without waiting for an answer (message 3). Finally, Chat_App_B informs User_B that the session has been closed by the other user.

3.3.1.7. *Send data*

The last important case is that of data sending, illustrated in Figure 3.10. In this case, we have only illustrated a case where one user sends data (User_A). The first message is sent by User_A (message 1). After that, three scenarios are possible: data is successfully sent and delivered (scenario "a"), data cannot be delivered at the receiver's side (scenario "b") or data cannot be sent to the receiver application (scenario "c").

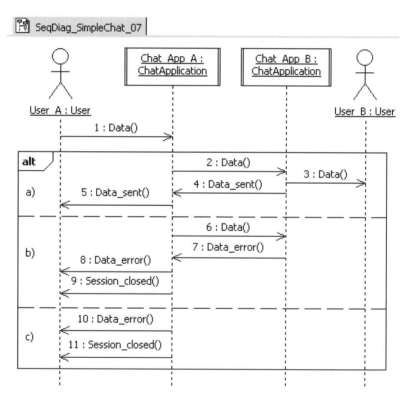

Figure 3.10. *Simple chat. Expected sequence diagram for "send data"*

In scenario "a", the sender application transmits the message through the network to the receiver application (message 2); after that, on receipt of the data message, the receiver application delivers the message to its user (message 3) and informs the sender application about the successful data reception (message 4). Finally, the sender application informs its user about the successful operation (message 5).

Scenario "b" represents the case where User_B is closing the session when data is received by Chat_App_B. In this scenario, the data is sent through the network to the receiver application (message 6). However, the receiver application is not able to deliver the data to its user anymore; thus, it informs the sender application that the session is being closed (message 7). Finally, the sender application informs its user about the error (message 8) and then it informs its user that the session has been closed (message 9).

Scenario "c" represents the case where the session is closed in both applications; in spite of this situation, User_A tries to send data through his application. In this scenario, Chat_App_A answers its user by indicating that there is an error (message 10) and that the session is closed (message 11).

3.3.2. *Concerned classes*

We have represented the possible interactions between a Chat Application and users. Let us now define the classes representing them.

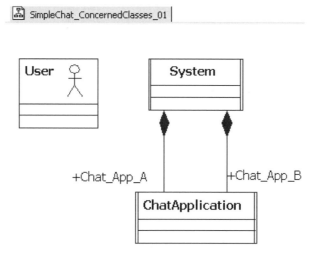

Figure 3.11. *Simple chat. Concerned classes*

As we have seen in section 3.3.1.1, there are two main classes: user and ChatApplication. Each class has two instances: User_A and User_B of type user, and Chat_App_A and Chat_App_B of type ChatApplication.

We will use a third class to represent the system and which will contain both instances of ChatApplication. Figure 3.11 represents these classes and their relationships.

3.3.3. *Signal list definition*

Based on the requirements specification and on the expected sequence diagrams described in section 3.3.1, it is possible to define a set of messages exchanged between the users and their application. These messages are represented in Figure 3.12[3].

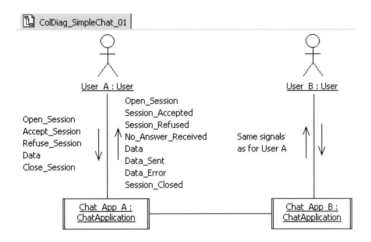

Figure 3.12. *Simple chat. Messages list between user and application*

Figure 3.13 describes the messages exchanged between the two instances of the ChatApplication class.

As you can see in Figure 3.13, all the messages exchanged between the ChatApplication instances are used as both inputs and as outputs at the same time.

---

3 The signals described above do not have parameters (actually, the only signal containing a parameter is "data"). The corresponding parameters will be added later in the design phase.

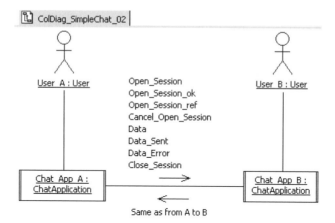

Figure 3.13. *Simple chat. Messages list between application instances*

Let us gather all the messages into interfaces. Figure 3.14 shows a class diagram containing the described interfaces.

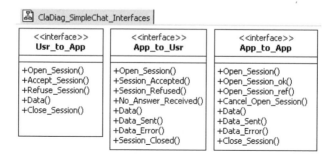

Figure 3.14. *Simple chat. Interfaces and messages*

We can now link the interfaces to their corresponding classes. For this, we first add some ports to the classes defined in section 3.3.2.

In Figure 3.15, you can see that we have added port P_U2App to the user class. This port sends and receives messages from a user to ChatApplication. We have linked Usr_to_App interface to this port as an output interface. Then, we have linked App_to_Usr interface to P_U2App port as an input interface.

Finally, we have added two ports to the ChatApplication class: P_App2U and P_App2App. The first port is used to communicate with a user, and the second to communicate with the other instances of ChatApplication class.

We have linked port P_App2U to App_to_Usr as an output interface and to Usr_to_App as an input interface. Then, we have linked P_App2App to App_to_App, as both input and output interfaces.

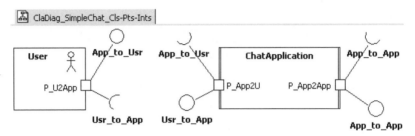

Figure 3.15. *Simple chat. Classes, ports and interfaces*

3.4. Architecture design

As noted in section 3.3.2, the system class is composed of two instances of the ChatApplication class: Chat_App_A and Chat_App_B. Thus, this is the system class that will define how its parts communicate between themselves.

It is also the system class which will receive signals from the users, pass them to the ChatApplication instances and then resend the signals from the ChatApplication to the users. We will represent this communication structure through a composite structure diagram.

In Figure 3.16, we have added two service ports to the system class: Port2Usr_A and Port2Usr_B. Remember that a service port only transfers received messages to an internal part. It also delivers messages coming from an internal part to an external element.

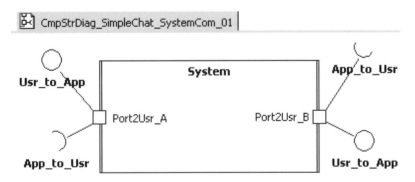

Figure 3.16. *Simple chat. Adding ports and interfaces to a system class*

After that, we have linked APP_to_Usr as an output interface and Usr_to_App as an input interface to both ports.

These two ports take all the messages coming from the users and relay them to the ChatApplication instances. They also take all the messages coming from the internal objects and relay them to the external users.

In Figure 3.17 we have represented the two system's internal parts: Chat_App_A and Chat_App_B.

Thus, we have defined the communication channels through which the messages will arrive from a system's ports to the chat applications and vice versa. We have called CH_Usr_APP_A the channel communicating to User_A, and CH_Usr_APP_B the channel communicating to User_B. Note that Port2Usr_A port is connected to P_App2U port in Chat_App_A. Reciprocally, Port2Usr_B is connected to P_App2U port in Chat_App_B.

Finally, we have added another channel in order to interconnect Chat_App_A and Chat_App_B.

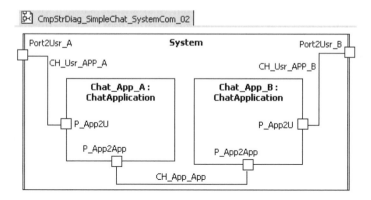

Figure 3.17. *Simple chat. Composite structure diagram for a system class*

3.5. Detailed design

We can now define the behavior of the ChatApplication class. To do this, we create a state machine managing the expected behavior for each scenario defined in the sequence diagrams described before.

We have to take into account that, since the system class is only a container, then it has no real internal behavior, and this class will only initiate its internal objects.

To begin with, a ChatApplication class is in a state where the session is closed and is waiting for an external stimulus. Figure 3.18 represents the initial state of this class.

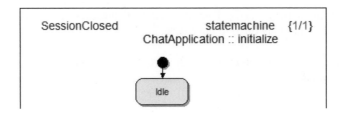

Figure 3.18. *Simple chat. Initial state machine for a ChatApplication class*

3.5.1. *Open session*

In the sequence diagram shown in section 3.3.1.2, it was defined that the first action is the receipt of an Open_Session message from a user.

In this scenario, when ChatApplication is in an idle state, it can receive two messages: either the Open_Session message from its user (Usr_to_App interface) or the Open_Session message coming from the other application (App_to_App interface). In order to allow us to understand the model, let us say that User_A starts the session (through Chat_App_A) and that User_B closes the session (through Chat_App_B).

Figure 3.19 shows the first option. In this case, on receipt of the Open_session message from the user, the chat application sends an Open_session message to the other application. After that, it starts a timer and waits for an answer from the partner application. We represent this waiting situation through a state that we have named openingSession.

Note that we are using a state machine diagram and that that there are three actions: a receive signal action, a send signal action and a regular simple action. Note also that the received and the sent signals have the same name: Open_Session; however, they are defined in different interfaces: Usr_to_App and App_to_App. Finally, note that we are using a timer inside the regular simple action. This regular simple action (represented by a simple square) sets Timer1 to NOW + ten time units. Every modeling system has a different method of managing time but this is how TAU G2 manages it.

For more details on the way state machine diagram graphic nodes are defined, see section 15.4, Diagrams, in [SSU 10].

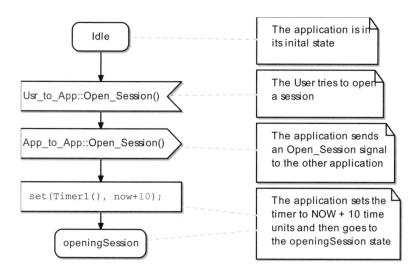

Figure 3.19. *Simple chat. State machine managing a session request from a user*

Figure 3.20 shows the second possibility in the Open Session scenario. In this case, on receipt of the Open_Session message from the partner application, the system asks its own user to either accept or refuse the session. We represent this by sending a message to the user and then going to a state where the system waits either for an answer from its user or for a cancellation from the other application.

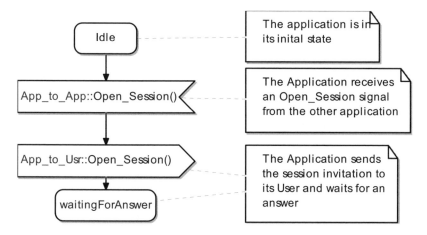

Figure 3.20. *Simple chat. State machine: session request from the second application*

3.5.2. *Timeout*

If no answer is received from the receiving side after a given time, the system on the requesting side (the side sending the request) will cancel the request and inform its user. The system then goes back to its initial idle state, in which it can receive a new request from its user or from the partner application.

In Figure 3.21, we start from the previously defined openingSession state. In this case, we see the timer interruption as a receive signal action. We can then see that two signals are sent, the first directed to the partner application in order to cancel the request, and the second directed to its user in order to inform him that no answer was received and that the request was cancelled. The system then goes into its initial idle state.

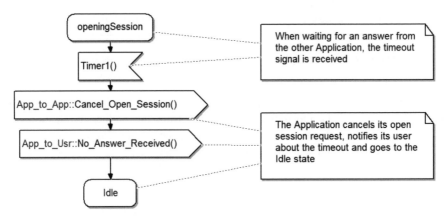

Figure 3.21. *Simple chat. State machine for the Timeout signal*

Now, we can model the expected behavior under the same scenario on side B. Figure 3.22 shows that, in this case, we start from the waitingForAnswer state. In this scenario, the only message to be received is Cancel_Open_Session. On receipt of this message, the system goes to its initial state.

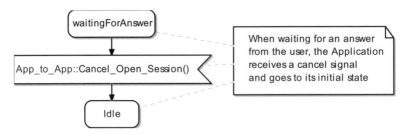

Figure 3.22. *Simple chat. Receiving session request cancellation*

3.5.3. Refusing the session

If the session invitation is refused by the requested user, then the application notifies the user making the request and then goes to its initial state.

Let us model first the expected behavior on side B. In Figure 3.23 we can see that the state machine starts from the waitingForAnswer state. In this case, on receipt of the Refuse_Session message from the user, the system sends an Open_Session_REF message to the partner application and then it goes to its initial state.

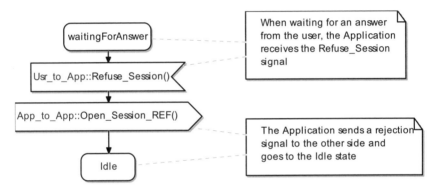

Figure 3.23. *Simple chat. Refusing a session request*

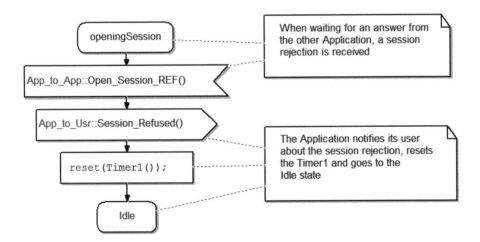

Figure 3.24. *Simple chat. Receiving a session request refusal*

Now, we can model the behavior on side A. In Figure 3.24 we see that the state machine starts from the openingSession state. In this case, on receipt of an Open_Session_REF message, the system sends a message to its user. Next, it resets the timer set in the previous scenario, and then it goes to its initial state. You can verify that this behavior is congruent with the expected behavior defined in Figure 3.2 in section 3.3.

3.5.4. *Accepting session*

After receiving the Open_Session signal from the user, application A is in its openingSession state, and application B is in its waitingForAnswer state. First, we will model what we expect to happen on the requested side when accepting a session request.

Figure 3.25 shows that the scenario starts from the waitingFromAnswer state. In this case, on receipt of the Accept_Session message from the user, the system sends an Open_Session_OK message to the requester application and then goes to another state where it will be able to manage the communication. We have called this state "sessionOpened". Note that the requesting side will be in the same state at the end of this scenario.

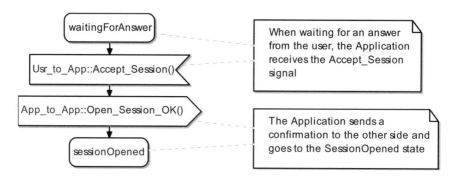

Figure 3.25. *Simple chat. Accepting a session invitation*

Now, let us model the behavior on side A. In Figure 3.26 we can see that this state machine starts from the openingSession state. On receipt of the Opening_Session_OK message from the requested application, the system sends an information message to its user. Next, the system resets the previously set timer and then goes to the sessionOpened state where it will be able to manage the data sending.

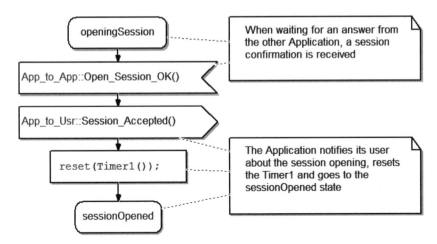

Figure 3.26. *Simple chat. Receiving a session acceptation indication*

3.5.5. *Closing session*

Let us imagine a simple closing scenario. Both applications are in the sessionOpened state and ChatApplication receives a Close_Session signal from its user.

In Figure 3.27 we can see that, on receipt of the Close_Session message from the user, the system sends a Close_Session to the partner application. Next, the system informs its own user and then goes to the initial state where it will be able to manage a new session request. As you would expect, the partner application will react in a similar way.

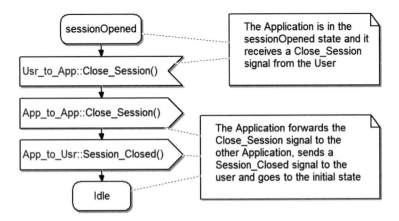

Figure 3.27. *Simple chat. State machine for initiating the session closing*

Figure 3.28 shows that on the counterpart application side, on receipt of a Close_Session message from the partner application, the system informs its user and then it goes back to its initial state.

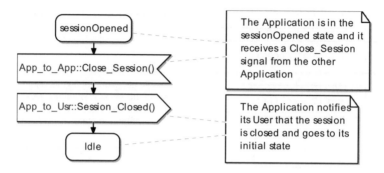

Figure 3.28. *Simple chat. Finishing the session closing process*

3.5.6. *Sending data*

The sequence diagram given in section 3.3.1.7 describes the expected behavior of the system when sending data. This scenario is triggered by the receipt of a Data message from the user. We can imagine three possible cases on receipt of this message from the user:

– case 1: successful delivery;

– case 2: unsuccessful delivery when the receiving user closes the session;

– case 3: unsuccessful delivery when the whole session is closed.

The difference is in the states of both chat applications when the data is sent. We will explain these three cases separately below.

3.5.6.1. *Case 1: Successful delivery*

In this case, both applications are in the SessionOpened state. Let us model first the behavior of the application receiving the Data signal from its user (App A).

In Figure 3.29 we can see that on the receipt of a data message from the user, the system will transfer the data received to the partner application and then wait for an acknowledgement from App B. Then, on receipt of the Data_Sent message from the partner application, the system will notify its user and go back to the SessionOpened state, where it will be able to process more data.

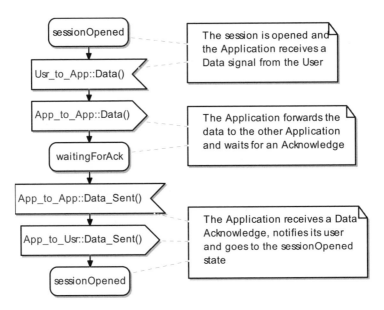

Figure 3.29. *Simple chat. State machine for a correct data delivery*

Figure 3.30 shows that, at the receiver side, on receipt of a data message from the partner application, the system presents the received data to its user. After that, the system informs the sender application that the data was correctly received and then it goes back to the sessionOpened state.

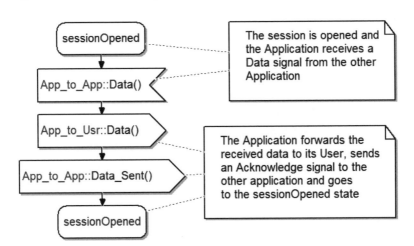

Figure 3.30. *Simple chat. A data signal is received and accepted*

As you can see, we are not dealing here with message losses, since we consider that that kind of problem would be managed by a lower level protocol. A protocol capable of handling errors will be modeled in a later chapter. Therefore, in this exercise, we consider that every sent message is correctly received and that every time the application is waiting for an answer, the answer is always received.

Note that, since we are not dealing with transport problems, we will only be dealing with functional errors.

3.5.6.2. *Case 2: the receiving user closes the session*

Remember that this case deals with the following scenario: the sender user sends data to his application (message 1 in Figure 3.31). In the meantime, User B closes the session. The receiver application is then in its idle state when it receives the message from the sender application (message 2). Moreover, the sender application is in its waitingForAck state when it receives the Data_Error message.

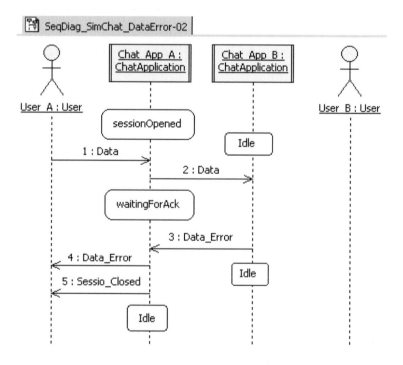

Figure 3.31. *Simple chat. The receiver application is in its idle state when receiving data from the partner application*

For this reason, *App_to_App::Data* messages cannot be processed by the receiver application (message 2 in Figure 3.31). We have solved this problem by sending an error message back to the sender application (message 3). After that, the sender application informs its user about the error (messages 4 and 5). This behavior corresponds to that expected, as defined in section 3.3.1.7.

In Figure 3.32 we can see the state machine managing this case on the receiver side. As we have said, the receiver application is in its idle state when it receives the Data message from the sender application. On receipt of this message, the system answers with a Data_Error message and then it goes back to its idle state. Remember that we have illustrated this behavior in Figure 3.10 in section 3.3.1.7.

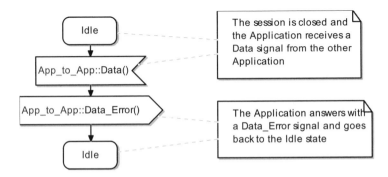

Figure 3.32. *Simple chat. A data signal cannot be received if the session is closed*

On the sender side, the first part of the behavior is the same as that given in the successful delivery scenario (see Figure 3.33).

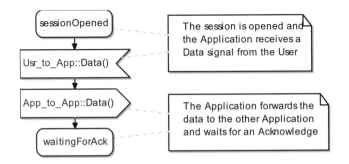

Figure 3.33. *Simple chat. First part of the successful delivery scenario*

In Figure 3.34, we see that, when the sender application is in its waitingForAck state, if it receives a Data_Error message from the receiver application, then the system informs its user that an error has occurred while sending the data. Next, it informs its user that the session is closed, and then goes into its idle initial state.

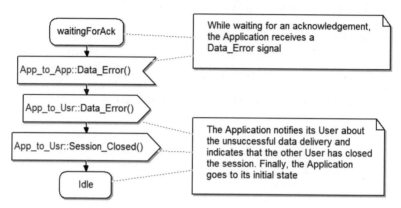

Figure 3.34. *Simple chat. The application realizes that the receiver has closed its session*

3.5.6.3. *Case 3: The whole session is closed*

This third case represents the error produced when the session is closed on sender and receiver sides and, in spite of this, the user tries to send data.

Figure 3.35 shows that the sender application is in its idle initial state when it receives a data message from the user. The application will then return a Data_Error message to the user. It will then inform its user that the session is closed, and then go back to its idle initial state.

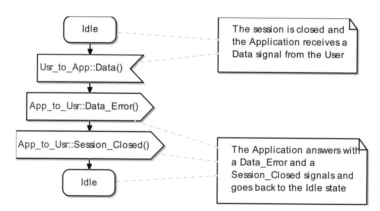

Figure 3.35. *Simple chat. A user tries to send data when the session is closed*

3.6. Simple chat simulation

3.6.1. *Intuitive test*

We can now validate the proposed model by simulation. This simulation produces a set of sequence diagrams that we can compare with the expected ones proposed in the analysis phase.

As shown in Figure 3.36, when we launch the first simulation, we obtain a very simple sequence diagram where both chat applications are in their idle state waiting for an input from the users, represented here by Port2Usr.

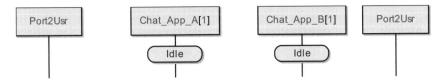

Figure 3.36. *Simple chat. Simulation: both applications go to their initial state*

Now, in Figure 3.37 we send an Open_Session message to Chat_App_A and we let the simulation continue automatically until a stable state is reached.

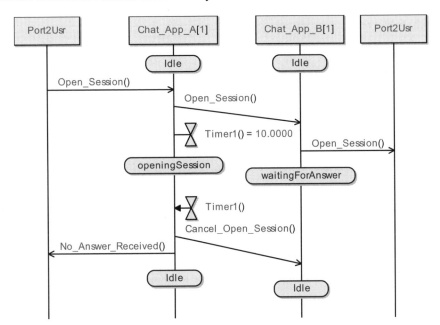

Figure 3.37. *Simple chat. Simulation: simulating the Open Session process with two users*

In this simulation, on receipt of the Open_Session message from the user, the requester application sends an Open_Session message to Chat_App_B; and then it launches a timer. Note the hourglass symbol used by TAU to represent the timer set. Then, we see that Chat_App_B asks its user to accept or to refuse the session. Note too that, at this point, the requester application is in its openingSession state while the requested application is in its waitingForAnswer state. This first part of the scenario was defined in section 3.5.1 and correctly matches the expected sequence diagram defined in section 3.3.1.2.

After that, since no answer is received by Chat_App_A, the timer expires. Note the hourglass with an arrow going to the lifeline; this symbol is used by TAU in order to represent a timeout interruption.

After the timeout signal reception, Chat_App_A cancels the request by sending a Cancel_Open_Session message to the partner application, and then it informs its user about the unsuccessful operation.

Note that both applications go back to their idle state where they will be able to manage a new request from their users. This second part of the scenario was defined in section 3.5.2 and correctly matches the expected sequence diagram defined in section 3.3.1.3.

3.6.2. *Initial tests*

We have validated, in an intuitive manner, the system behavior when a user opens a session and no answer is received (open session and timeout scenarios defined in sections 3.3.1.2 and 3.3.1.3).

We want now to validate the entire system behavior. To do so, we will follow the scenarios defined in the analysis phase. Remember that we described these scenarios by using sequence diagrams. These scenarios represent a coherent and logical set of actions within a normal system utilization. These scenarios represent most of the operational behaviors.

As a last step, we will test the system behavior when a non-coherent sequence of messages is injected, for example:

– sending a close session message when the session is already closed;

– sending an open session message when the session is already opened;

– sending a data message before receiving the acknowledge message, etc.

We will test the system in the same order that it was analyzed.

3.6.2.1. *Refuse session*

In Figure 3.38 we can see the sequence of messages representing the Open_Session scenario.

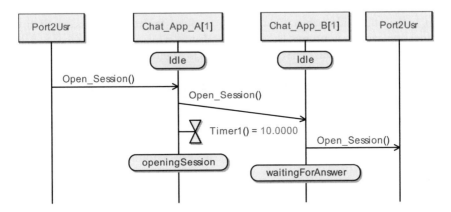

Figure 3.38. *Simple chat. Simulation: sending the Open_Session signal to User B*

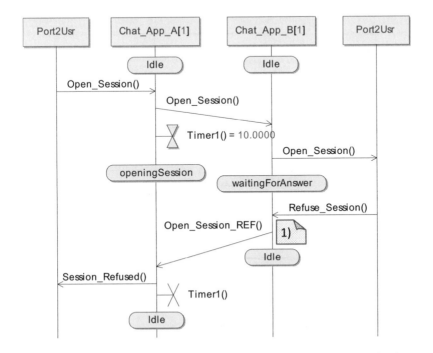

Figure 3.39. *Simple chat. Simulation: simulation of the Open Session refusal*

After that, in Figure 3.39, User_B (represented by Port2User on the right) refuses the session (note (1)). On receipt of the refusal message from the user, Chat_App_B sends a refusal message to the requester application, and then goes to its idle state. This behavior was defined in Figure 3.23 in section 3.5.3.

On the other side, when Chat_App_A receives the refusal message from the partner application, it informs its user. After that, it resets its internal timer and then goes to its idle state. This behavior was defined in Figure 3.24 in section 3.5.3.

We note here that the sequence diagram obtained by simulation correctly matches the expected behavior defined in section 3.3.1.4.

3.6.2.2. *Accept session*

After the request for a session opening, we can see in Figure 3.40 that User_B accepts the session (note (1)).

On receipt of the acceptance message from the user, Chat_App_B sends an acceptance response to the requester application and then goes to its sessionOpened state. This first part of the scenario was defined in Figure 3.25 in section 3.5.4.

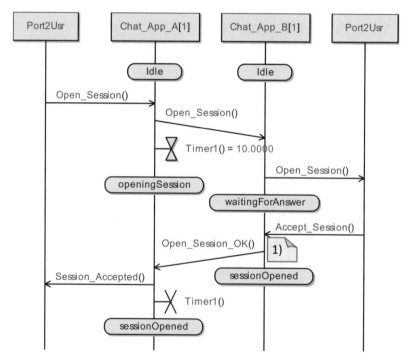

Figure 3.40. *Simple chat. Simulation: accepting the Open Session request*

On receipt of the OK answer from its partner application, Chat_App_A informs its user. After that, Chat_App_A resets its internal timer and then goes to its sessionOpened state. This second part of the scenario was defined in Figure 3.26 in section 3.5.4.

Observe that this sequence diagram obtained by simulation correctly matches the one defined in section 3.3.1.5.

3.6.2.3. *Close session*

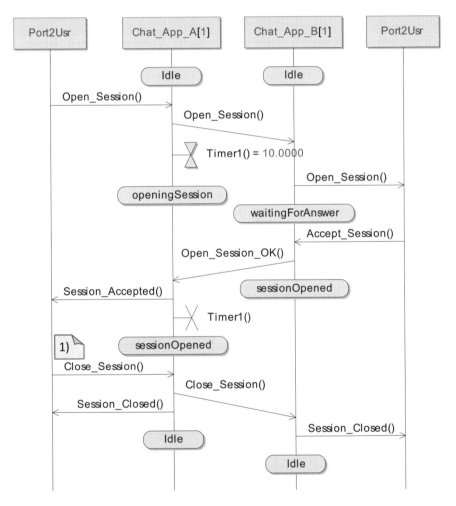

Figure 3.41. *Simple chat. Simulation: closing session*

Figure 3.41 shows that, after opening the session, User_A (represented here by Port2User on the left) closes the session (see note (1)). On receipt of this message, the Chat_App_A sends a Close_Session message to its partner application. After that, the requester application sends a Session_Closed message back to its user. Finally, it goes back to its idle state. This behavior was described in Figure 3.27 in section 3.5.5.

On the User_B side, on receipt of the Close_Session message, the Chat_App_B application informs its user, then goes back to its idle state. This behavior was defined in Figure 3.28 in section 3.5.5.

Note that this simulation correctly matches the expected behavior defined in section 3.3.1.6.

3.6.2.4. *Send data*

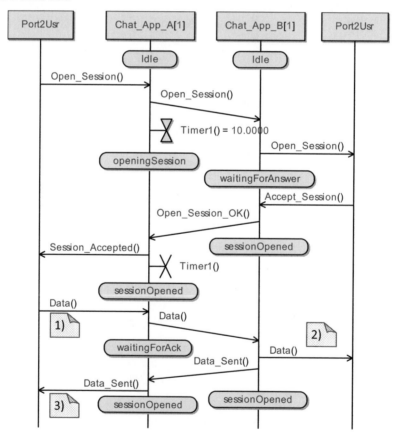

Figure 3.42. *Simple chat. Simulation: data sending simulation, part 1*

In Figure 3.42 we see that, after opening the session, User_A sends a data message to its application (note (1)). On receipt of this message, Chat_App_A transfers the received data to its partner application and then goes to its waitingForAck state. After receipt of the confirmation, Chat_App_A informs its user and then goes to its sessionOpened state (note (3)). The same behavior is observed when User_B sends a data message (see Figure 3.43, notes (4) and (6)). This behavior was defined in Figure 3.29 in section 3.5.6.1.

On the receiver side, when Chat_App_B receives a data message from its partner application, it informs its user. Then it goes back to its sessionOpened state (see Figure 3.42, note (2)). The same behavior is observed on the User_A side (see Figure 3.43, note (5)).

This behavior was formalized in Figure 3.30 in section 3.5.6.1

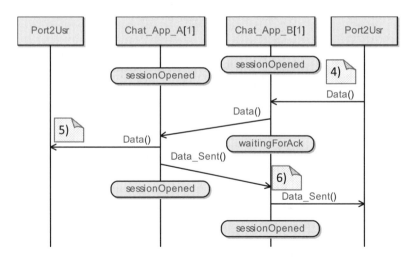

Figure 3.43. *Simple chat. Simulation: data sending simulation, part 2*

Again, this sequence diagram correctly matches the expected behavior defined in section 3.3.1.7.

This simulation only covers a fraction of the main "send data" scenario. Remember that the send data process has three possible outcomes, one of them, successful data delivery, was validated through this simulation. The other two cases correspond to error cases. We will test them in section 3.6.3.

3.6.2.5. *Coverage statistics*

At this point, we have tested our model through many simple scenarios. Each simulation covers a certain number of states and transitions. For example, the simulation described in section 3.6.1 (open session request, then time out) only covers 26% of states and 29% of transitions (see Figure 3.44).

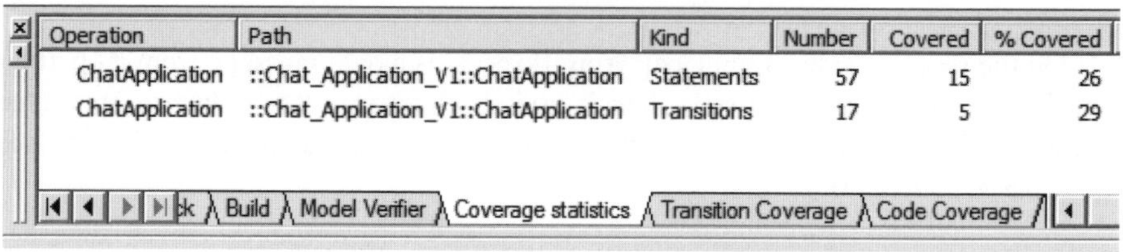

Operation	Path	Kind	Number	Covered	% Covered
ChatApplication	::Chat_Application_V1::ChatApplication	Statements	57	15	26
ChatApplication	::Chat_Application_V1::ChatApplication	Transitions	17	5	29

|◄| ◄ | ► | ►| k \ Build \ Model Verifier \ Coverage statistics \ Transition Coverage \ Code Coverage \|| ◄ |

Figure 3.44. *Simple chat. Coverage statistics:*
open session request, timeout

The simulation described in section 3.6.2.4 (open session request, accept session, User_A sends data and User_B sends data) only covers 45% of states and 47% of transitions (see Figure 3.45).

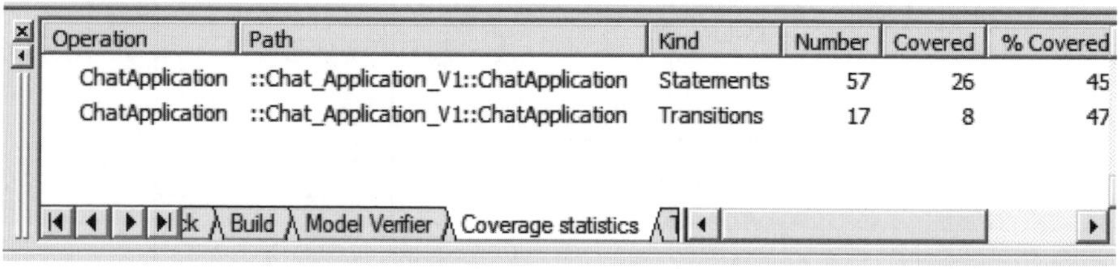

Operation	Path	Kind	Number	Covered	% Covered
ChatApplication	::Chat_Application_V1::ChatApplication	Statements	57	26	45
ChatApplication	::Chat_Application_V1::ChatApplication	Transitions	17	8	47

|◄| ◄ | ► | ►| k \ Build \ Model Verifier \ Coverage statistics \ ◄ |

Figure 3.45. *Simple chat. Coverage statistics: open session, accept, send data*

However, we do not know how many states and transitions were tested in total, and so we do not know how many states and transitions are still to be tested. A possible solution could be to view all the transitions and all the states tested on each simulation and gather them into a single set. However, this solution is quite long and tedious. For example, if we analyze the list of states for the two previous coverage statistics, we obtain the lists shown in Figures 3.46 and 3.47.

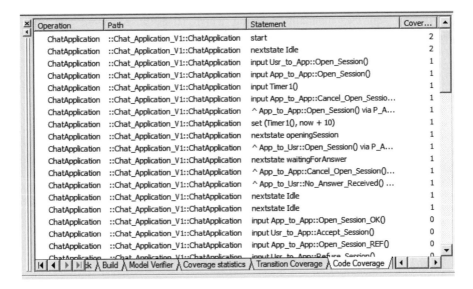

Figure 3.46. *Simple chat. Coverage statistics: list of covered states*
(open session and timeout)

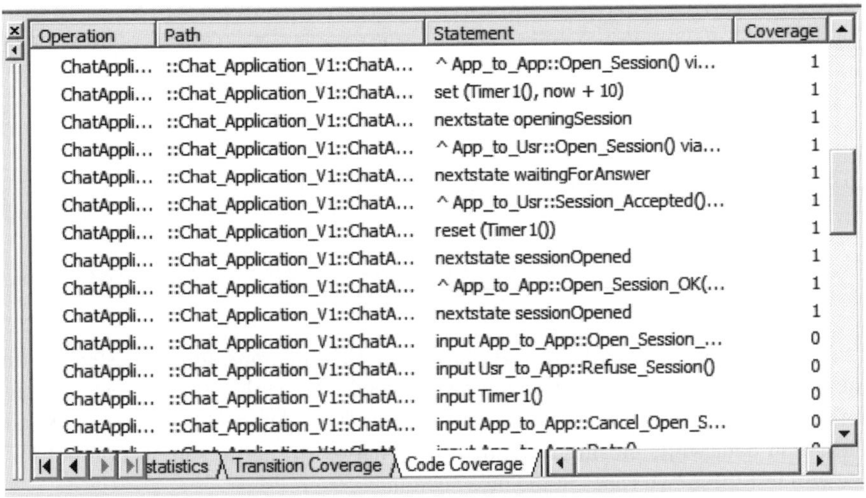

Figure 3.47. *Simple chat. Coverage statistics: list of covered states*
(open session, accept and send data)

As you can see, manually deciding which states were visited and which transitions were fired on two separate simulations might be a very difficult and error prone task.

An easy way to solve this problem is to define a big scenario containing all the simple scenarios previously defined, for example in sequence:

– *User_A* opens a session and no answer is received from *User_B*;

– *User_A* opens a session;

– *User_B* refuses the session;

– *User_A* opens a session;

– *User_B* accepts the session;

– *User_A* sends data;

– *User_B* sends data;

– *User_A* closes the session.

If we perform this sequence of actions, the coverage statistics indicate that 80% of states and 82% of transitions were covered (see Figure 3.48)

Operation	Path	Kind	Number	Covered	% Covered
ChatApplication	::Chat_Application_V1::ChatApplication	Statements	57	46	80
ChatApplication	::Chat_Application_V1::ChatApplication	Transitions	17	14	82

|◄|◄|►|►|ild \ Model Verifier \ Coverage statistics \ Transitio|◄| |

Figure 3.48. *Simple chat. Coverage statistics: multiple scenarios together*

Note that there are still 20% of states and 18% of transitions untested by our simulation. This can be easily explained. Indeed, the explanation is that we have defined two cases that have not yet been tested: "send data when session is closed" and "send data while closing" (defined in scenarios (b) and (c) in section 3.3.1.7). We will test these two cases in the next section.

3.6.3. *Extended tests (fault tolerance)*

In previous sections, we have tested our model on coherent signal sequences, i.e. open – timeout, open – refuse, open – accept – send – close. In this section, we will test a set of non-coherent sequence of signals, for example: open – open, send without opening session, open – accept – close – close, etc.

Modern UML modeling systems allow the designer to verify the percentage of transitions and states that have been visited after a simulation but, for the moment, they do not permit automatically generating the tests. As a consequence, defining the tests is the responsibility of the software designers, and in particular they need to define a set of simulations containing incoherent scenarios. These scenarios allow the designer to test the system behavior under unexpected but possible (maybe due to error) signal sequences.

3.6.3.1. *Proposed scenarios*

As an example, the following set of scenarios enables potential analysis and design errors to be found. For each test, if the simulation reveals any error, the model has to be correctly redesigned.

This list is not exhaustive, but it is representative of the most common errors made during protocol design (unexpected or incorrect signals, simultaneous signal sending, etc):

– *A-Sends data when session is closed.* One of the users sends data without taking into account that the session is closed.

– *A-Sends data while B is closing the session.* One of the users sends data. At the same time, the other user starts closing the session.

– *Close (with session closed).* When a user tries to close an already closed session, it has to be notified about the session status.

– *Open – Open.* When a user tries to open an already opened session, it has to be notified about the session status.

– *A-Opens – B-Opens.* Both users open the session at the same time.

– *A-Closes – B-Closes.* Both users close the session at the same time.

– *A-Sends – B-Sends.* Both users send data at the same time. The system has to deliver the first received data without losing the second, and then deliver the second data.

3.6.3.2. *Send when session is closed*

When a session is closed, both applications are in their Idle state (see Figure 3.49). In this scenario, User A sends data to his application. Then, the chat application answers with a Data_Error message. After that, the application informs the user that the session is closed, and it goes back to its idle state. This behavior was defined in section 3.5.6.3.

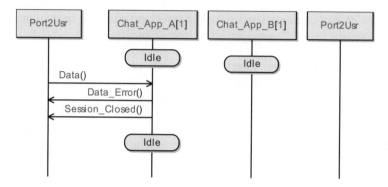

Figure 3.49. *Simple chat. Simulation: send with session closed*

This sequence diagram correctly matches the expected behavior defined in section 3.3.1.7.

3.6.3.3. *A-Sends while B-Closes*

Figure 3.50 gives the resulting sequence diagram for this scenario.

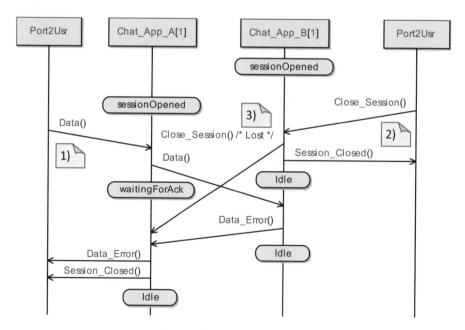

Figure 3.50. *Simple chat. Simulation: a user sends data while the other closes the session*

Before analyzing the obtained sequence diagram, let us take a moment in order to verify the coverage statistics since we have tested the two missing scenarios. If we execute all the scenarios viewed at this point together, we obtain the statistics described in Figure 3.51. Note that 100% of states have been visited and that 100% of transitions have been triggered by our simulations.

We could expect that no error is present in our model.

Operation	Path	Kind	Number	Covered	% Covered
ChatApplication	::Chat_Application_V1::ChatApplication	Statements	57	57	100
ChatApplication	::Chat_Application_V1::ChatApplication	Transitions	17	17	100

Figure 3.51. *Simple chat. Coverage statistics: adding the two missing scenarios*

We can now go back to the analysis of the sequence diagram obtained by simulation. In Figure 3.50 we see that the session is open. Hence, both applications are in their sessionOpened states.

In note (1), we can see that User A sends data to his application. At that moment, in note (2), we see that User B sends a Close_Session message to his application. After that, both applications send a message to their respective partner application. In note (3), we observe that the Close_Session signal from Chat_App_B to Chat_App_A is lost.

We should now carefully analyze the rest of the messages. First, after Chat_Application_B sends the Close_Session signal to Chat_Application_A, it confirms the action to its user (Session_Closed message) and then goes to its initial state. This means that it is not expecting to receive any other data message from its partner application. Fortunately, we previewed this situation in section 3.3.1.7 and defined the corresponding behavior in section 3.5.6.2. Thus, in this case, Chat_App_B answers with a Data_Error message.

On the User A side, the situation is in fact different. Indeed, in section 3.5.6.2, we anticipated the receipt of the Data_Error message when the application is in its waitingForAck state. However, we did not anticipate the receipt of the Close_Session message when it is in this state. This is the reason why the Close_Session message between the applications is lost; i.e. from the current state, there is no transition able to read the received message. If we want to define a logically safe simulation without lost messages (i.e, without leaving unprocessed messages in the buffers), then we need to manage the receipt of the Close_Session message in the waitingForAck state.

In order to avoid this loss of the message, we need to accept the Close_Session signal with no action, and we will wait for the Data_Error signal. The state machine described in Figure 3.52 implements this new behavior.

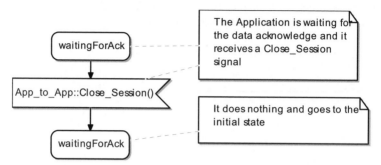

Figure 3.52. *Simple chat. Managing the Close_Session signal while sending data*

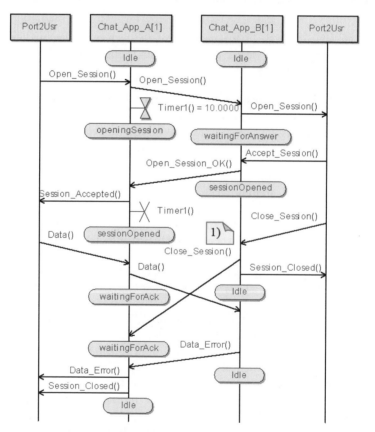

Figure 3.53. *Simple chat. Simulation: "close session" message loss corrected*

We can now repeat the same scenario and see if the problem persists.

Figure 3.53 shows the resulting sequence diagram. In note (1), observe that the Close_Session message is now correctly managed; the Chat_App_A reads the message and ignores it, and then it goes back to the waitingForAck state.

Figure 3.54 shows the new coverage statistics for a complete simulation containing the new behavior. Again, note that 100% of transitions have been executed and that 100% of states have been visited.

Operation	Path	Kind	Number	Covered	% Covered
ChatApplication	::Chat_Application_V2::ChatApplication	Statements	59	59	100
ChatApplication	::Chat_Application_V2::ChatApplication	Transitions	18	18	100

◄ ◄ ► ►◄ Coverage statis ◄

Figure 3.54. *Simple chat. Coverage statistics: "close session" message loss corrected*

This example allows us to understand that, even if all the states have been visited and all the transitions have been triggered by the given set of simulations, this cannot fully validate the system behavior. Undeniably, some functional errors might persist since some data values or signal sequences have not been simulated. The following tests are aimed at finding and correcting these kinds of functional errors.

3.6.3.4. *Close (with session closed)*

Figure 3.55 shows the sequence diagram obtained from the simulation when a user tries to close an already closed session. Not that the Close_Session message is lost. In other words, the application is not prepared to receive this message while being in the idle state: therefore, it ignores the message.

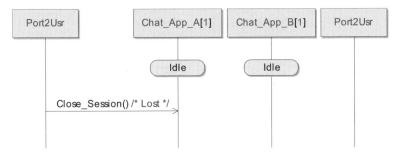

Figure 3.55. *Simple chat. Simulation: "close session" message is lost*

This kind of behavior is misleading because it is preferable to explicitly notify the user that the session is closed. Let us modify the ChatApplication behavior.

The ChatApplication can currently manage the following messages when it is in an idle state:

– Open_Session from user;

– Open_Session from application;

– data from user;

– data from application.

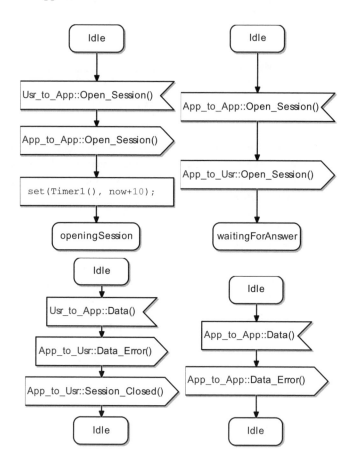

Figure 3.56. *Simple chat. Reminder of messages managed from the idle state*

The state machines representing this behavior are described in Figures 3.19 and 3.20 in section 3.5.1, Figure 3.32 in section 3.5.6.2 and Figure 3.35 in section 3.5.6.3. Figure 3.56 recalls these state machines.

We can note here that ChatApplication cannot receive any other signal while being in the idle state. This is why any other signal is currently lost. Now, in order to avoid this kind of behavior, the application will be modified in order to read and process all other signals.

Let us add the following behavior to our initial requirements:

– when ChatApplication is in the Idle state, it will answer any message sent by the user which is different from Data or Open_Session with the Session_Closed message.

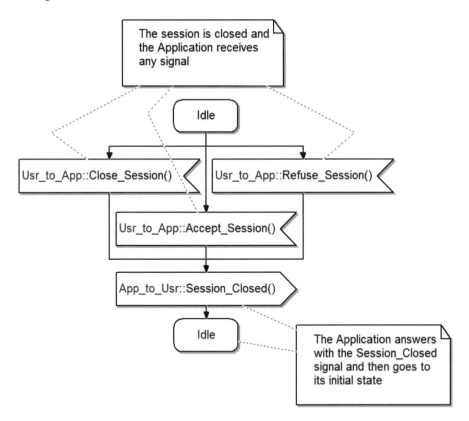

Figure 3.57. *Simple Chat. Managing unexpected signals when the session is closed*

Figure 3.57 allows ChatApplication to manage this new behavior. As we know, there are only five possible signals from user to application (see section 3.3.3) and two of them are already processed. As a consequence, this new state machine ensures the management of the three other signals: Close_Session, Accept_Session and Refuse_Session.

Let us try again to send a Close_Session signal when the session is closed. Note now that, in Figure 3.58, ChatApplication does not ignore the message and answers it with a Session_Closed signal.

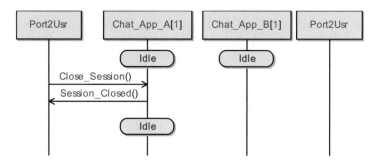

Figure 3.58. *Simple chat. Simulation: "close session" message is correctly managed*

Figure 3.59 shows the behavior obtained when the user sends the Accept_Session and Refuse_Session messages. Observe that no message is lost.

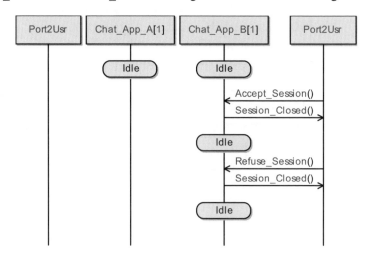

Figure 3.59. *Simple chat. Simulation: ChatApplication manages the two missing messages on a closed session*

3.6.3.5. *Open (when session is already opened)*

Figure 3.60 shows the sequence diagram obtained when a user tries to open an already opened session. As you can see in note (1), the second Open_Session message is lost. This is because the application does not expect to receive this kind of message while being in the sessionOpened state.

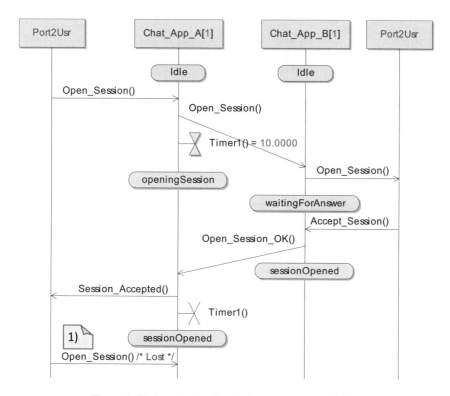

Figure 3.60. *Simple chat. Simulation: open session is lost*

In order to correct this problem, we can add the following behavior:

– when the user tries to open an already opened session, the system will answer with a Session_Opened message.

This new behavior implies adding a new message to the model. Figure 3.61 shows the new interfaces. This diagram substitutes the one presented in Figure 3.14, in section 3.3.3.

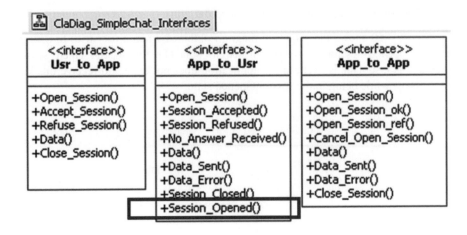

Figure 3.61. *Simple chat. Adding the Session_Opened message to the interface list*

Figure 3.62 shows the state machine managing this new behavior. Note that this state machine starts when the system is in the sessionOpened state. On receipt of the Open_Session message from the user, it answers with the Session_Opened message and then it goes back to the sessionOpened state.

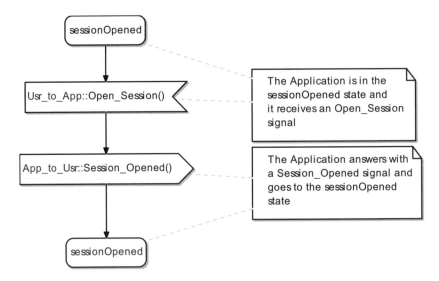

Figure 3.62. *Simple chat. Validating unexpected signals when the session is opened*

Let us consider the new simulation. We can see at note (1) in Figure 3.63 that, this time, the message is not lost and the system correctly answers the request.

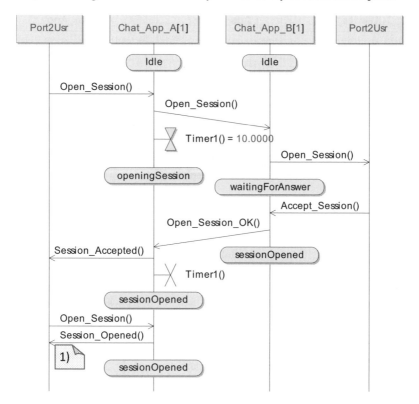

Figure 3.63. *Simple chat. Simulation: the system correctly answers the open request when the session is already opened*

Nevertheless, the system might receive two other messages from the user while being in the sessionOpened state: Accept_Session and Refuse_Session. We can anticipate that both messages will be lost.

In order to avoid this incorrect behavior, we can add the following requirement:

– when the user tries to accept or to refuse a session when a session is currently opened, the system will answer with an Illegal_Signal message.

We can then update the list of messages within the interfaces once more (see Figure 3.64).

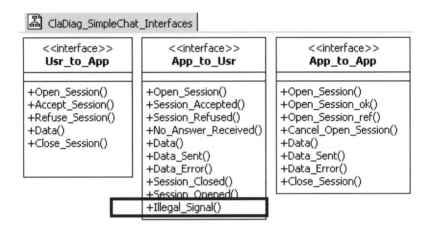

Figure 3.64. *Simple chat. Updating the list of messages in the interface list*

Figure 3.65 shows the state machine managing this new situation. Observe that the state machine starts from the sessionOpened state. On receipt of the Accept_Session or of the Refuse_Session messages from the user, the system answers with an Illegal_signal message. Then, it goes back to the sessionOpened state.

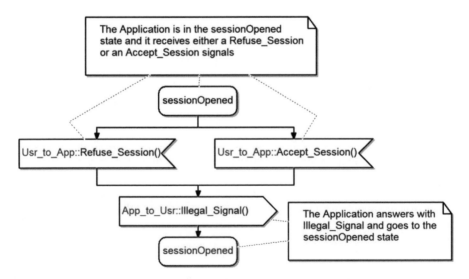

Figure 3.65. *Simple chat. Validating unexpected signals when the session is opened*

Now, we can run a new simulation in order to test this new behavior. Observe in notes (1) and (2) in Figure 3.66 that the chat application correctly answers this new sequence of messages from the user.

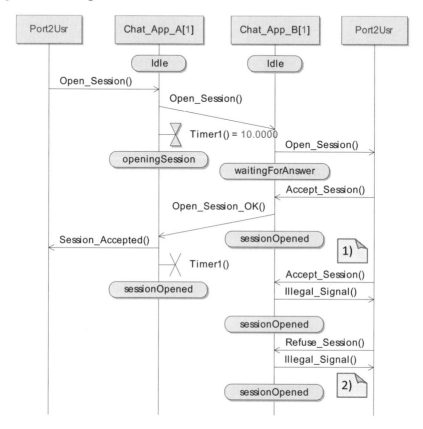

Figure 3.66. *Simple chat. Simulation: the system correctly manages two more messages when the session is opened*

3.6.3.6. *A-Opens – B-Opens*

In this scenario, we check how the system reacts when both users try to open a session at the same time. Figure 3.67 gives the sequence diagram obtained after simulation. Observe (at notes (1) and (2)) that both Open_Session messages sent from one application to the other are lost. The reason for this is that both applications are in their openingSession state and are waiting for Open_Session_OK, Open_Session_REF or timeout signals. This means that the application is not ready to receive the Open_Session message in this state, and the message is then lost.

Now, look at notes (3) and (4). The Cancel_Open_Session messages are also lost. This is because the applications are in their idle state when they receive this message and do not expect to receive it.

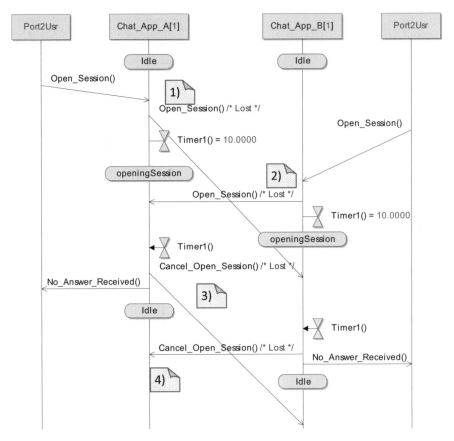

Figure 3.67. *Simple chat. Simulation: error found when both users try to open the session at the same time*

A possible solution to this problem is to consider the Open_Session message to be an acknowledgement of the session opening request. Indeed, the receipt of an Open_Session message in the openingSession state means that the partner application is also trying to open the session.

The state machine given in Figure 3.68 starts in the openingSession state. Then, on receipt of the Open_Session message from the partner application, it informs its user that the session has been established. After that, the application resets its internal timer, and then goes to the sessionOpened state.

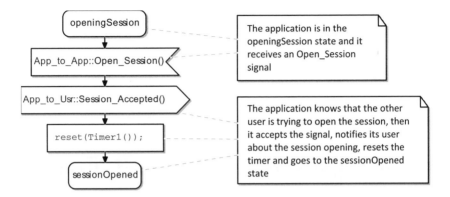

Figure 3.68. *Simple chat. Correcting the error on simultaneous open session requests*

Figure 3.69 shows the new sequence diagram obtained when both users try to open a session at the same time.

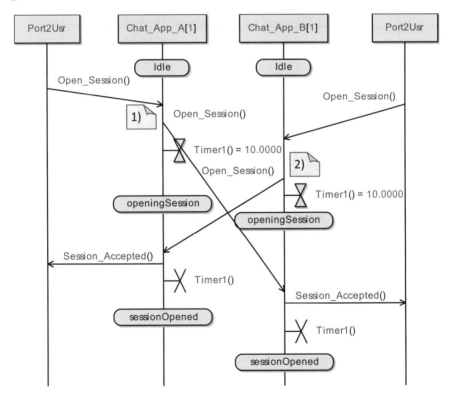

Figure 3.69. *Simple chat. Simulation: accepting simultaneous open session requests*

Observe in notes 1 and 2 that, this time, the Open_Session messages are not lost: in fact, they are considered to be a confirmation to the request and the session is successfully established.

3.6.3.7. A-Closes – B-Closes

Once the session is opened, if both users try to close the session at the same time, the current model might lead to an unexpected behavior, as shown in Figure 3.70.

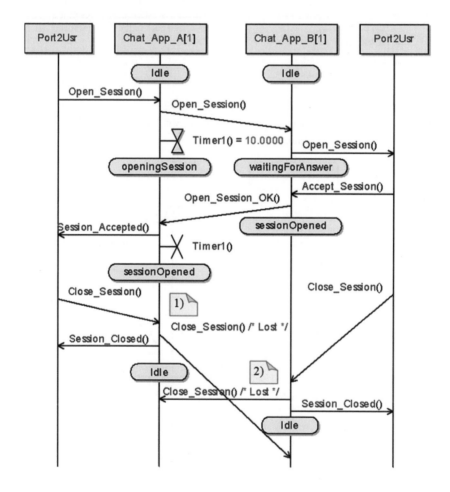

Figure 3.70. *Simple chat. Simulation: error found when both users close the session at the same time*

As you can see, the simulation loses the Close_Session messages between the applications. Even if this loss does not change the result, it is a design error that should be corrected.

A possible solution to this error consists of specifying that the application is able to receive a Close_Session message from its partner application with no action. The state machine shown in Figure 3.71 implements this new behavior.

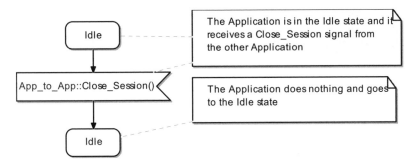

Figure 3.71. *Simple chat. Correcting the Close–Close error*

Figure 3.72 shows the new sequence diagram obtained from the simulation of the new state machine. As you will notice, it appears that the Close_Session messages between the applications are accepted, although with no consequences to the current system state.

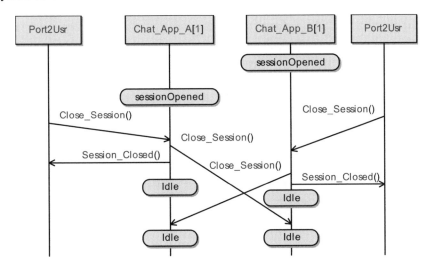

Figure 3.72. *Simple chat. Simulation: simultaneous close is accepted*

3.6.3.8. *A-sends – B-sends*

One very interesting scenario is as follows: how will the system react if both users send data at the same time?

Figure 3.73 shows the sequence diagram describing the system behavior when facing this scenario.

As you can see in notes (1) and (2), the data messages are lost. In fact, both of the applications are in their waitingForAck state and are waiting for a Data_Sent or a Data_Error signal; consequently, they are not able to receive a Data message.

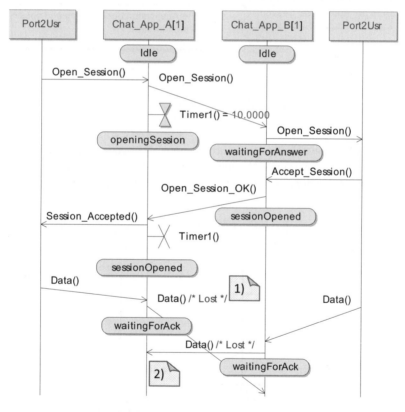

Figure 3.73. *Simple chat. Simulation: error when both users send data simultaneously*

A more serious problem is revealed by this scenario: both of the applications are frozen in their waitingForAck state. This shows that they are not able to receive any other message from the users, i.e. data, Close_Session, etc.

We can correct this problem by accepting a data signal while being in the waitingForAck state. After that, the system will go back to the waitingForAck state and will wait for a Data_Sent or a Data_Error signal.

Figure 3.74 shows the state machine implementing this new behavior. Starting from the waitingForAck state, and on receipt of a data message, the system informs its user about the received data. After that, the system acknowledges the received data and then goes to the waitingForAck state again.

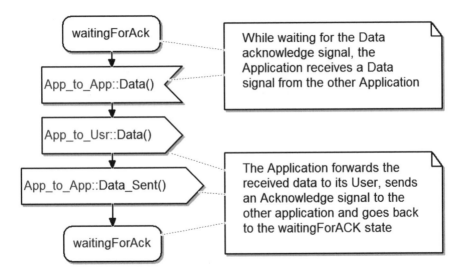

Figure 3.74. *Correcting the send–send error*

Figure 3.75 shows the new sequence diagram obtained from the simulation of this new behavior. In this figure, you can observe at note (1) that the data messages between the applications are not lost anymore. Then, at notes (2) and (3) you can see that the received data is delivered to the user. Finally, at note (4) you can see that the received data are acknowledged by the Data_Sent message.

Other scenarios are possible, for example, what happens if one of the applications goes down suddenly? In that case, the still active instance would not know if the other was still alive or not, and then could wait indefinitely for an answer. This problem can be solved by using a "heartbeat" to test whether the other instance is still alive or not. Another solution might be to define a limited time value for each expected answer.

Another interesting question might follow from this last solution: if we do not know if the delay is caused by an application error, by network congestion or a network problem, what is the correct waiting time?

We will not give the solution to this problem here; however, it is a good exercise for the reader to think about and to propose a feasible solution.

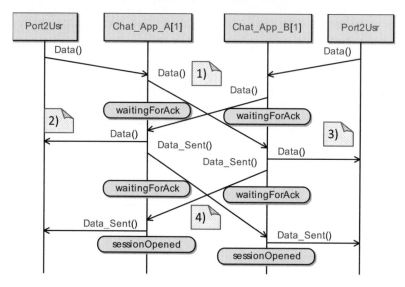

Figure 3.75. *Simple chat. The system correctly manages a simultaneous data send*

3.7. Chapter summary

This chapter presented a simple chat application. This application was specified and modeled.

We then validated the model behavior and some fault tolerance scenarios were tested.

Finally, we proposed the changes necessary in order to correct the errors found during the verification phase.

This chat application allows us to start thinking about communicating systems from an object-oriented perspective. It also gives us a very good insight into the potential offered to designers by simulations, since it is possible to detect and correct algorithmic and functional problems at an early design phase.

3.8. Bibliography

[SSU 10] Unified Modeling Language™ (UML®), Superstructure specification, 2010-05-05, http://www.omg.org/spec/UML/2.3/

Chapter 4

Non-reliable Transmission Mediums

4.1. Introduction

Chapter 3 presented a simple chat application, but that application assumes the existence of a perfect transmission medium, i.e. a medium in which there are neither losses nor unexpected delays or message disordering. However, this is not currently the case with the Internet.

Nowadays, applications use the TCP (Transmission Control Protocol) [RFC 93], a protocol to ensure the correct reception of data. However, this solution is not applicable for real-time multimedia data[1]. Most applications dealing with this kind of problem implement their own solutions but another possibility is the creation of a generic transport protocol that will make these problems transparent for all the applications.

Below, we propose a simple transport model and protocol to solve the losses and delay problem generated in the medium. To do so, we will divide our model into two parts:

1. The definition of an imperfect transport medium. This transport medium can lose and/or delay transmitted packets[2]. Let us take into account that the packet disordering problem is a natural consequence of a variable delay. This medium can be used to test if the proposed protocol behaves correctly in such cases. In particular,

---

1 The explanation of why this protocol is not applicable for real-time multimedia data does not lie within the scope of the book; however, suffice to say, the time deadline required for such multimedia data can be missed.

2 This medium could be seen as the UDP (User Datagram Protocol) [RFC 68].

the medium must be dynamically configured, i.e. the defined loss rate and maximal delay limit should be able to be dynamically defined by a supervising user.

2. The proposal of a transport protocol solving the problem of losing and delaying messages. This protocol will offer a reliable service to an application.

This chapter presents the model of the Non-Reliable Medium, while the next chapter will present the simple transport model and protocol.

4.2. Requirements

Two remote entities[3] want to communicate through a medium application.

– The main task of the medium is to transfer data from one entity to another.

– The medium provides the means to be configured by an external Controller.

– The transmitted data is transparent for the medium, i.e. the medium works in the same manner, no matter what type of data transmitted.

– The medium applies two different mechanisms to each data packet before delivering it: packet losing and packet delaying. To take these in turn:

– Packet losing:

- on receipt of a data packet, the medium applies a random loss mechanism in order to decide if the data packet is to be delivered or not to the receiver entity;

- the medium manages a losing rate parameter;

- the medium sets the default losing rate to 0% (no losses);

- the medium manages a different losing rate in both directions;

- the medium allows the Controller to dynamically configure the losing rate parameter for each direction;

- the minimum and maximum values of the losing rate parameter are 0 and 100%, respectively.

– Packet delaying:

- on receipt of a data packet, the medium applies a random delay mechanism before delivering it to the receiver entity;

- the medium manages a random delay between 0 and MAX_DELAY time-units;

- the medium allows the Controller to dynamically configure the MAX_DELAY;

---

3 An entity here represents an application or a protocol.

- MAX_DELAY will never be lower than 0;

- the medium sets the default MAX_DELAY to 0 (no delay);

- the medium manages a different MAX_DELAY for each direction.

Note that the requirements presented here do not fully represent a real network, which is more complex, but permit a didactic example to be defined[4].

4.3. Analysis

We can represent the given requirements as a scenario modeled with an activity diagram[5].

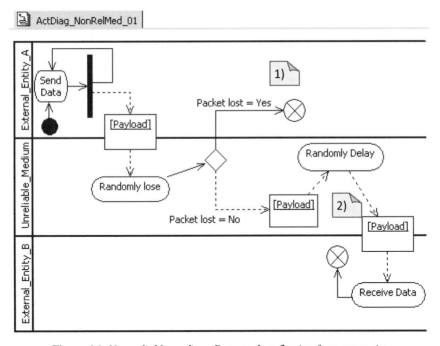

Figure 4.1. *Non-reliable medium. Data packets flowing from one entity to another through the medium*

---

4 A real network model should take into account many other features, for example, congestion and disorder.
5 This activity diagram uses Control Flows, Object Flows, Fork Nodes, Object Nodes, Actions and Flow Final nodes. For more details on these artifacts, see Chapter 12, "Activities", in [SSU 10].

In Figure 4.1 we see that, thanks to the fork node, the External_Entity_A sends a data packet (represented by a Payload object node) to the Unreliable_Medium and, simultaneously, the control flow goes back to the Send_Data action. This organization allows an entity to send data without waiting for a message to be delivered.

We can follow the object flow going to Unreliable_Medium and observe that the Unreliable_Medium randomly loses the packet. If the packet is effectively lost, then the control flow goes to the FlowFinal node (see note (1)). Otherwise, if the packet is not lost, then the medium randomly delays the packet (see note (2)). Finally, the data packet is delivered to the External_Entity_B and the control flow goes to the FlowFinal node.

As you can imagine, a symmetrical flow is followed by the data packets sent by the External_Entity_B.

Figure 4.2 shows a complete representation of the data sending process. We can see that, thanks to the first fork node, both entities can send data simultaneously. We can also observe that the data packets coming from any entity are processed in the same way by the medium.

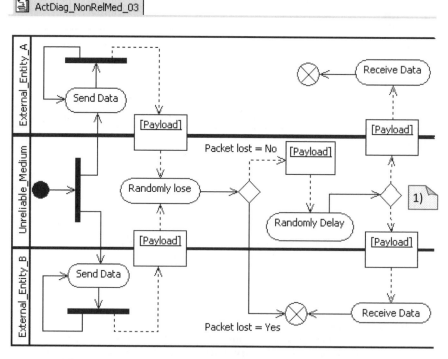

Figure 4.2. *Non-reliable medium. Symmetrical flow of data packets*

Nevertheless, this diagram presents a weakness. Indeed, the current organization does not allow us to decide to which entity the data packets will be delivered at the end of the process (see note (1)).

A second problem arising in this diagram concerns the requirement indicating that the loss rate and the maximal delay might be configured separately for each flow. In this diagram, both flows are treated the same way.

In fact, the current organization has to be extended to be able to differentiate each flow in order to decide which parameters should be applied to each packet.

In order to alleviate this, we have decided to define a different control for each of the flows, as can be seen in Figure 4.3. Observe that each flow follows its own path. This configuration allows us to define a different loss and a different delay configuration for each direction in an easy way.

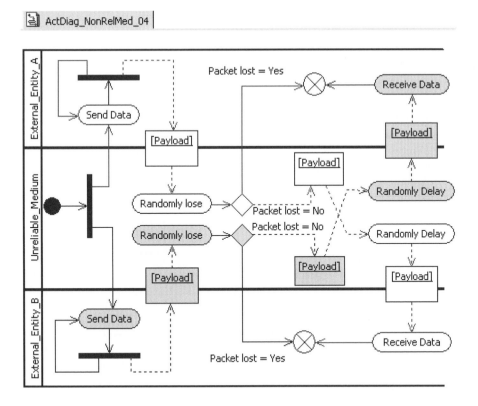

Figure 4.3. *Non-reliable medium. A different path for each data flow*

Note that this model also allows us to avoid the problem of deciding to which entity the data packets will be delivered at the end of the process. Certainly, in this organization the packets going from A to B always follow the same path, and this path is different from the path followed by the packets going from B to A.

Finally, note that we did not represent the configuration process in our diagrams since we consider that the sequence of actions required to configure the medium is quite simple. However, we will take the configuration into account in the following sections.

4.3.1. *Sequence diagrams*

In order to create the state machines representing the behavior of the non-reliable medium we need the list of messages exchanged between the entities and the system under study. We also need to know at what moment the internal operations are called.

The solution to this task is to create a sequence diagram based on the activity diagram described in Figure 4.3. Let us present the expected behavior.

4.3.1.1. *Configure*

Figure 4.4 represents the expected behavior for the configure scenario. As you can see in message 1, the controller sends a configure message to the NR_Medium. If the received configuration is incorrect, then the NR_Medium returns an ErrorMSG message to the controller. Otherwise, no answer is sent to the controller.

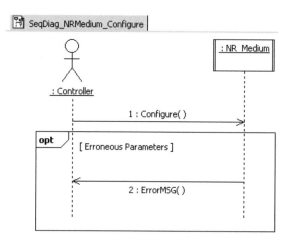

Figure 4.4. *Non-reliable medium. Expected sequence diagram for configuration*

4.3.1.2. *Send*

Figure 4.5 shows the expected behavior for the send scenario. Application_A[6] sends a Data message to the NR_Medium. You will certainly have noted that the data message has no parameter and no payload. For simplicity's sake, we have decided to ignore the payload in this phase of the analysis. Since no treatment is performed by the medium (remember that the payload is transparent to the medium), we will add the payload parameter at the end of our modeling work.

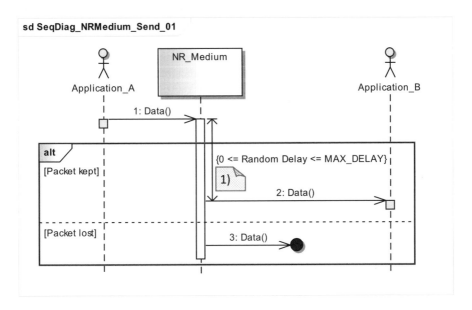

Figure 4.5. *Non-reliable medium. Expected sequence diagram for send data (01)*

On receipt of the data message, two alternatives are possible; either the packet is delivered or it is lost. In the first case, when the packet is delivered, a random delay is applied to it.

We represent this random delay by a "duration constraint" between messages 1 and 2[7]. This duration constraint indicates that there is a time duration higher than or equal to 0 and lower than or equal to MAX_DELAY between data reception and data resend by the medium (see note (1) in Figure 4.5). After the random delay is applied, the data packet is delivered to Application_B by the medium.

---

6 Remember that, in practice, a non-reliable medium can receive data from any entity, i.e. from an application, a protocol, a user or even another medium.
7 For more details on time constraints, see Chapter 14 in [SSU 10].

In the second case, the packet is lost by the medium. We represent this loss through the lost message node (see message 3).

This diagram correctly represents our requirements. However, we cannot see where the internal processes (packet loss and random delay) occur. Let us therefore refine the diagram.

Figure 4.6 shows that, if the packet is not lost, then an internal timer is set by the NR_Medium (see message 2). We represent this timer behavior by a "self" message. After that, the system waits for a timer interruption. The sequence diagrams are not well suited to represent the system interruptions. As a consequence, we represent the timeout using a self message again. Even if the representation is not very accurate, we focus on a simple graphical representation indicating that the system waits for a timeout interruption before resending the data packet.

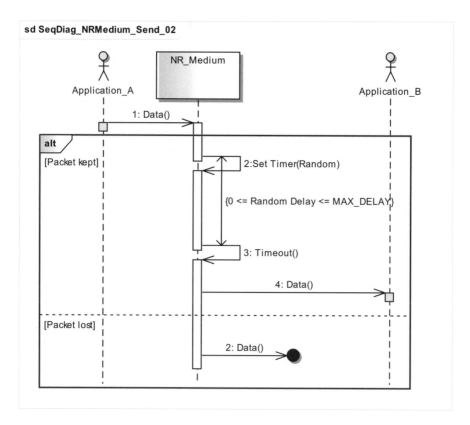

Figure 4.6. *Non-reliable medium. Expected sequence diagram for send data (02)*

Note that random loss is represented by the "alt" combined fragment. We can imagine that on receipt of a data packet from an application, the NR_Medium will apply a Random_Loss process which will decide if the received packet will be kept or lost. We can devine that the Random_Loss process occurs just after the data reception and before the Random_Delay occurs.

4.3.2. *Concerned classes*

As explained before, in order to facilitate the data packet processing, we define a different path for packets coming from each direction.

An easy solution to this is to define two internal classes within the medium: losing and delaying. These two classes will be instantiated and separately configured by the medium.

By following our analysis in section 4.3, a data packet received by the medium is first passed to the losing process and then to the delaying process, before being delivered. Figure 4.7 illustrates this behavior.

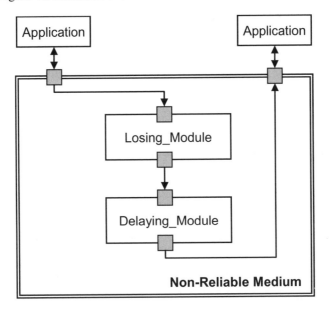

Figure 4.7. *Non-reliable medium. First view of the internal architecture*

As explained before, we could have used a single losing module and a single delaying module in order to process the data packets coming from both entities, as shown in Figure 4.8. However, this solution makes it more difficult to separately set the losing and delay parameters applied to each flow, and it complicates the delivery differentiation.

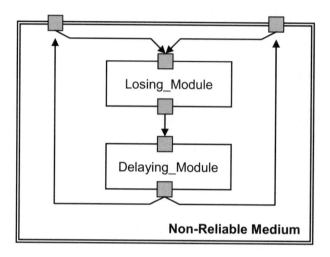

Figure 4.8. *Second view of the medium's internal architecture*

However, as explained in Figure 4.3 in section 4.3, we have decided to use a simpler solution. Our model will duplicate the losing and delaying modules once for each direction, as shown in Figure 4.9.

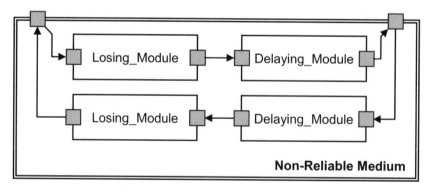

Figure 4.9. *Internal architecture for the non-reliable medium*

4.3.2.1. *Accessory classes*

We propose to add an accessory class called "random". This class will be used by the losing and the delaying classes, and will generate the needed random numbers, of two types:

– a random number between 0 and 1. This number will be used by the losing module in order to decide if the received data packet will be delivered or not;

– a random number between A and B. This number will be used by the delaying module in order to assign a random delay between the configured limits to each data packet.

4.3.2.2. Class diagram

At this point, we have considered four important classes and they are represented in Figure 4.10. NR_Medium is the main class and it represents the non-reliable medium itself. This class is composed of two instances of delaying class, Delaying_AB and Delaying_BA. It is also composed of two instances of losing class, Losing_AB and Losing_BA. As you see, we have used the composition association in order to represent their relationships. Finally, the NR_Medium class uses the accessory Random class previously proposed. Note the simple navigable[8] association between the NR_Medium and the random class. This means that the Random class can be accessed by the NR_Medium class, while the inverse is not permitted[9]. Finally, note that the NR_Medium, losing and delaying classes are active. Remember that an active class has its own thread of control, while passive classes do not. Active classes are represented by a class box with an additional vertical bar on either side[10].

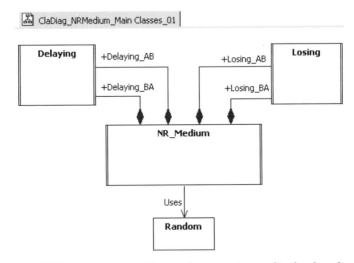

Figure 4.10. *The losing and the delaying classes are inserted in the class diagram*

---

8 The navigability is represented by an arrow head at the end of the association line.
9 For more details on different kind of associations, see sections 6.4.2, Diagram format, and 7.3.3, Association (from Kernel), in [SSU 10].
10 For more details on active and passive classes, see section 13.3.8 in [SSU 10].

4.3.3. *Signal list definition*

Figure 4.11 summarizes the list of messages exchanged between the medium and its applications[11]. As you will see, the list of messages is quite short. The application entities can only send and receive data messages, while the controller can send configuration messages and receive error notifications.

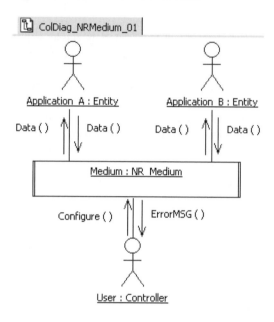

Figure 4.11. *Non-reliable medium. Messages exchanged between the medium and the applications and the user*

We can create the interfaces gathering the messages represented in the previous figure.

⊞ ClaDiag_NRMedium_Interfaces_01

<<interface>> **App_to_Medium**	<<interface>> **Medium_to_App**	<<interface>> **User_to_Medium**	<<interface>> **Medium_to_User**
+Data()	+Data()	+Configure()	+ErrorMsg()

Figure 4.12. *Non-reliable medium interfaces*

---

11 The parameters contained in the signals are not shown in this phase of the model.

We have defined four interfaces: two interfaces for communicating with the applications and two for communicating with the controller user.

The interfaces for communicating with the applications (App_to_Medium and Medium-to-App) contain a single message: data. The interfaces for communicating with the controller user are organized as follows: User_to_Medium contains the message Configure, while Medium_to_User contains ErrorMsg.

Again, in this phase of the modeling process, we do not need to represent the configuration parameters in the Configure message.

Now, we can link these interfaces to their corresponding classes, first adding ports to the NR_Medium class.

Observe in Figure 4.13, the ports added to this class: Port_M_A (port medium to application A), Port_M_B (port medium to application B) and Control_Port_M. You can also see that Port_M_B and Port_M_A use App_to_Medium as an input interface and Medium_to_App as an output interface, while Control_Port_M uses User_to_Medium as an input interface and Medium_to_User as an output interface.

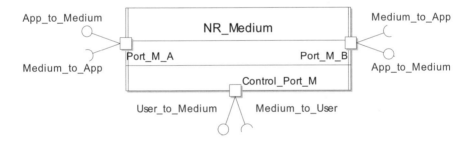

Figure 4.13. *Non-reliable medium. Linking the interfaces to main class*

As we have explained in Figure 4.9, in section 4.3, the losing class will receive the data messages from the application. Then, we add a port, Losing_In, to this class and link it to App_to_Medium as an input interface (Figure 4.14).

The delaying class has the responsibility of sending data to the applications. Then, we add a port Delaying_Out to the delaying class and link it to Medium_to_App as an output interface.

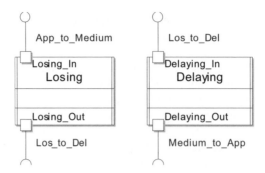

Figure 4.14. *Non-reliable medium. Linking the corresponding interfaces to losing and delaying classes*

On receipt of a data packet from an application, the losing class decides if the packet is transmitted to the receiver application or not. If so, then the losing class sends a message to the delaying class. For this reason, we have added ports Losing_Out and Delaying_In to the losing and delaying classes, respectively. We have also created a new interface, Losing_to_Del, and linked it to the Losing_Out and Delaying_In ports, as output and input interfaces, respectively. As you can imagine, the Los_to_Del interface contains a single message, data. Figure 4.15 shows the new list of interfaces.

🖥 ClaDiag_NRMedium_Interfaces_02

<<interface>> **App_to_Medium**	<<interface>> **Medium_to_App**	<<interface>> **User_to_Medium**
+Data()	+Data()	+Configure()

<<interface>> **Medium_to_User**	<<interface>> **Los_to_Del**
+ErrorMsg()	+Data()

Figure 4.15. *Non-reliable medium. New list of interfaces*

Figure 4.16 shows a class diagram gathering all the classes, their ports and the interfaces linked to those ports.

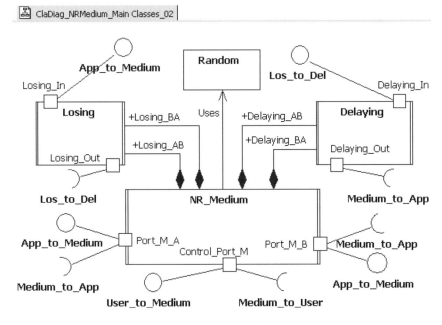

Figure 4.16. *Non-reliable medium. Complete diagram: classes, ports and interfaces*

4.4. Architecture design

We have defined the associations between all classes of our model. We have also added the ports needed to enable communication between the outside world and the inner elements, and between these inner elements. Finally, we have linked the interfaces to their corresponding ports.

Now, we will describe the communication paths between the NR_Medium class internal modules.

Take a look at Figure 4.17. You will see that the communication paths are clearly identifiable. Data packets coming from application A arrive at Port_M_A, and then are carried to the Losing_AB module through the chINab channel. After that, if the packet is not lost, then it is transported to the Delaying_AB module through the chLDab channel. Finally, after a random time, the packet is sent to the Port_M_B port through the chOUTab channel. At that time, the packet will be delivered to application B.

The communication path from application A to B is received by NR_Medium through Port_M_B port, and carried to Losing_BA through the chINba channel.

After that, if the packet is kept, then it is transferred to Delaying_BA through the chLDba channel. Finally, the packet is sent to application A through the chOUTba channel and Port_M_A port.

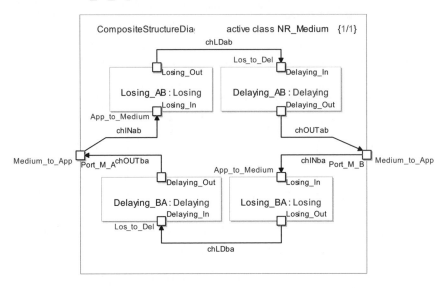

Figure 4.17. *Non-reliable medium. Internal architecture of the NR_Medium class*

4.4.1. *Detailing the sequence diagram*

The simulations that we will perform in the following sections will show the losing and delaying instances. Thus, we need to refine our previous sequence diagrams by adding the modules composing NR_Medium in order to better compare the simulation results against the expected behavior.

Figure 4.18 shows the path followed by a data packet going from application A to application B. We can see some little differences compared with Figure 4.6 in section 4.3.1.2. We can observe that it is the Losing_AB module which receives the data packet from the application. We can also see that, on receipt of this packet, this module performs an internal operation that we have called "randomly lose" (see message 2). This internal operation will decide whether the data packet will be lost or not.

After that, we can observe that, if the data packet is not lost, then it is transmitted to Delaying_AB (see message 3). We can also note that it is the Delaying_AB module which sets the timer, receives the timeout interruption and sends the packet to its final destination.

Finally, observe that, in the case when the packet is lost, it is Losing_AB which decides not to send the data packet to Delaying_AB.

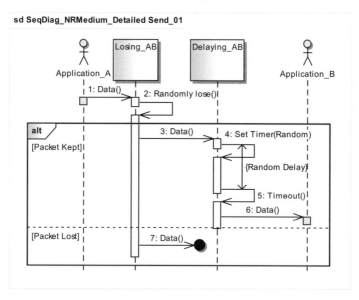

Figure 4.18. *Non-reliable medium. Detailed sequence diagram for sending data from A to B*

Of course, the data packets flowing in the opposite direction will follow a symmetrical path. This path is illustrated in Figure 4.19.

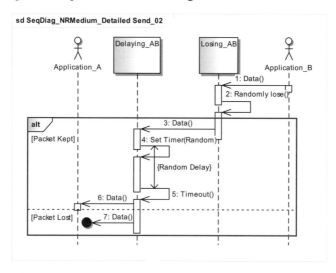

Figure 4.19. *Non-reliable medium. Detailed sequence diagram for sending data from B to A*

We are now also able to give more details about the configuration scenario.

Figure 4.20 shows the sequence diagram detailing the expected behavior for the configuration process. As before, the controller user sends a configure message to the medium, but this time we have added four parameters, one for each internal module. After that, if the received parameters are within the expected range, the medium configures each of its internal modules. We have represented this configuration as a function-call to an operation named Set defined for each module (see messages 2 to 5).

By contrast, if the received parameters are outside the expected range, then the medium returns an ErrorMsg message to the controller user (see message 6).

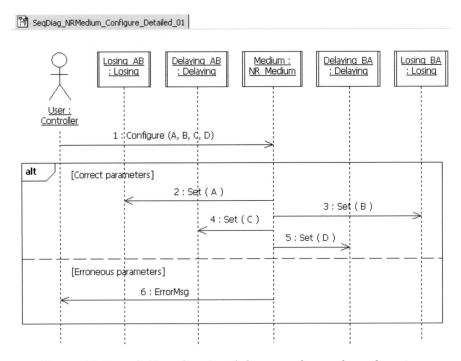

Figure 4.20. *Non-reliable medium. Detailed sequence diagram for configuration*

4.5. Detailed design

Now, we will not only use state machines or activity diagrams to define the classes' behavior but will also define some code. We begin, of course, with some easy operations and will then add more complexity in the next sections.

As you remember, we have been using Telelogic TAU G2 in order to simulate our models. In this chapter, we will take advantage of a few of the features existing in this software. For instance, it contains a text diagram. A text diagram can be used to describe the textual definition of a given element within an existing diagram; for example, it can be used to textually define an operation body. This text diagram uses a particular syntax based on C++ and Java but extended to cover non-programming language concepts, for example stereotypes and tagged values[12].

We will now detail the classes composing our model.

4.5.1. *Random class*

As explained before, this class will calculate random numbers. To do so, we use three operations:

– public void ini(): this operation initializes the seed value for random operations. This operation needs no parameters;

– public Real rand(): this operation returns a real random value between 0 and 1. This operation needs no parameters;

– public Real randAB(Real:A Real:B): this operation returns a real random value between A and B. This operation receives two real numbers defining the range.

Figure 4.21 shows the random class with its operations' declaration.

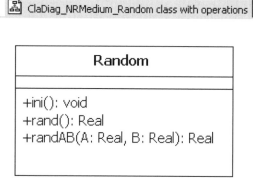

Figure 4.21. *Non-reliable medium. Random class with operations*

---

12 Details of the UML textual syntax used by TAU do not fall within the scope of this book. However, most of the modeling and simulation tools offer similar functionalities.

4.5.1.1. *ini operation*

In Figure 4.22 we see the content of the text diagram defining the *ini* operation body. The symbols "[[" and "]]" (lines 1 and 3) are used by TAU to insert C++ code directly into a text diagram. Line 2 runs the "srand" method contained in "cstdlib" or "stdlib" (see section 4.5.1.2 below for further explanation). This method reinitializes the C++ random generator with a given seed value (the local time in our example) in order to prevent random numbers from being the same every time the simulation is executed.

```
1.      [[
2.      srand(time(NULL));
3.      ]]
```

Figure 4.22. *Non-reliable medium. C++ code for the Random::ini operation*

4.5.1.2. *rand operation*

In Figure 4.23 we see the code defining the behavior of the rand operation. Note that line 1 declares a new variable, r1, of type real. Lines 2 and 4 define a C++ code block.

In order to better understand the operation body, more explanations are needed. In this operation, we are mixing the local UML textual syntax used by our simulation software and a native C++ code. A C++ bloc accesses a variable defined outside it by using the following syntax: #(name_of_the_variable). We then need to know that C++ defines a method, rand(), which returns a random integer number between 0 and RAND_MAX (RAND_MAX being a constant defined in <cstdlib>).

Now, in line 3, the variable, r1, is assigned with a random real number between 0 and 1. Finally, line 5 returns r1.

```
1.      Real r1;
2.      [[
3.      #(r1)=(float)rand()/((float)RAND_MAX);
4.      ]]
5.      return r1;
```

Figure 4.23. *Non-reliable medium. Code for the Random::rand operation*

4.5.1.3. *randAB operation*

In order to define this operation, we use a state chart diagram with an action symbol[13] containing code (see Figure 4.24). Remember that this operation receives two real parameters, A and B.

Line 1 defines a real variable, zero_one, which stores a random number between 0 and 1. In line 2 we define a real variable, ret, which stores the random number between A and B to be returned. Line 4 calculates a random number between A and B. Finally, in line 5, we return the required random number.

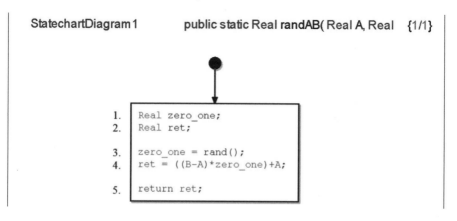

Figure 4.24. *Non-reliable medium. State machine for the Random::randAB operation*

4.5.2. *Losing class*

For this class, we use one private attribute and two operations:

– private integer LossesRate: this attribute stores the loss rate to be used by the class. The default value for this attribute is 0;

– public void setLossesRate(integer): this operation sets the value of the LossesRate attribute. It receives as parameter an integer value from 0 to 100;

– public void Initialize(): this operation is equivalent to the "main" function[14] in C++ programming. It will receive the signals from the application and randomly loses them.

---

13 For more details on action symbols and other graphic nodes on transitions, refer to section 15.3.14, Transition, in [SSU 10].
14 Remember that the "main" function is the point by where all C++ programs start their execution, independently of its location within the source code.

Figure 4.25 shows the Losing class with its attributes and operations declaration.

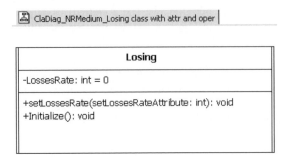

Figure 4.25. *Non-reliable medium. Losing class with attributes and operations*

4.5.2.1. *setLossesRate operation*

The definition of this operation is quite easy and we can use a text diagram to describe it. As you will see in Figure 4.26, the only action in this operation is to assign the received setLossesRateAttribute to the local private LossesRate attribute.

```
1.      LossesRate=setLossesRateAttribute;
```

Figure 4.26. *Non-reliable medium. Code for the Losing::setLossesRate operation*

4.5.2.2. *Initialize operation*

This operation is more interesting than the previous ones.

As you can see in note (1) in Figure 4.27, the first step is to declare an internal integer variable, called randomValue, which will store a random number. After that, the system goes to its idle state and waits for a data packet coming from the application (note (2)). On receipt of a data packet from the application (see note (3)), the system gets a random number between 0 and 99 by using the Random::randAB operation (see note (4)). Then, the system compares the obtained random number with the LossRate variable. If the random number obtained is lower than the defined rate, then the data packet is discarded (see note (6)). Otherwise, the data packet is sent to the corresponding delaying instance (see note (5)). In both cases, the system returns to its idle state.

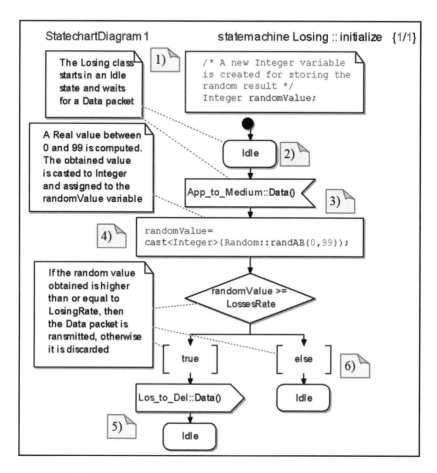

Figure 4.27. *Non-reliable medium. Code for the Losing::setLossesRate operation*

4.5.3. *Delaying class*

In order to implement this class, we use two attributes and two operations:

– private integer MAX_DELAY: this attribute specifies the maximal delay. The default value for this attribute is 0;

– private timer dataTimer: this timer controls the random delay. The system can set and start this timer, after which, the timer triggers an interruption when the defined time is over[15];

---

15 A full explanation of how the simulation software manages time is not within the scope of this book.

– public void setMaxDelay(integer): this operation sets the value of the MAX_DELAY attribute. It receives a non-negative integer value as a parameter;

– public void Initialize(): this operation is equivalent to the main function in C++ programming. It receives the signals from the losing module and randomly delays them.

Figure 4.28 shows the Delaying class with its attributes and operations.

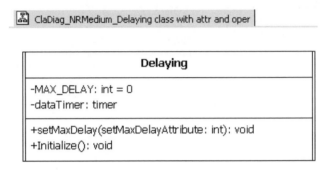

Figure 4.28. *Non-reliable medium. Delaying class with attributes and operations*

4.5.3.1. *setMaxDelay operation*

Again, the definition of this operation is quite easy and we can use a text diagram to describe it. As you can see in Figure 4.29, the only action in this operation is the assignment of the received setMaxDelayAttribute attribute to the local private MAX_DELAY attribute.

```
1.      MAX_DELAY = setMaxDelayAttribute;
```

Figure 4.29. *Non-reliable medium. Code for the Delaying::setMaxDelay operation*

4.5.3.2. *Initialize operation*

This operation needs two extra variables (see note (1) in Figure 4.30):

– an integer variable, called randomValue, to store the random values;

– a duration variable, called randomTime, needed to set the timer.

Indeed, our simulation software requires two parameters, one of type timer and one of type duration[16] in order to set a timer.

The system starts in its idle state waiting for a message from the losing module (see note (2)). On the receipt of a data packet, the system computes a random number between 0 and MAX_DELAY (see note (3)). After that, the random number is converted into duration. Then, the dataTimer variable is set to now + the obtained random number. Finally, the system returns to its idle state.

As you can see, this state machine defines and starts a timer with a random delay for a data packet.

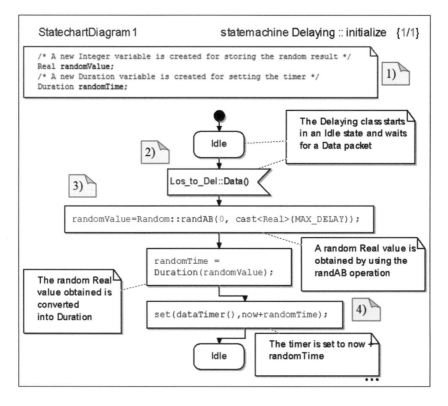

Figure 4.30. *Non-reliable medium. State machine for the initialize operation in the delaying class*

---

16 Duration is a data type used by our simulation software. Full details about this data type are not within the scope of this book.

Now, we need to read the timeout signal and send the data packet to the receiver application. In Figure 4.31, we see that the system is in its idle state. On receipt of a timeout interruption, the system sends the data packet and then goes back to its idle state.

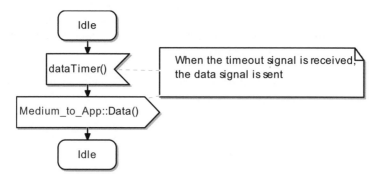

Figure 4.31. *Non-reliable medium. State machine for an initialize operation in the delaying class (continued)*

4.5.4. *Non-reliable medium class*

We have already defined the attributes of this class in section 4.3.2.2: *Losing_AB*, *Losing_BA*, *Delaying_AB* and *Delaying_AB*. Now we need an operation to launch the internal parts[17] and to configure them.

This behavior will be provided by a new operation named "public void Initialize()", which is equivalent to the main function in C++ programming. It receives the control signals from the user and configures its internal parts.

4.5.4.1. *Initialize operation*

You can see, in Figure 4.32, that this operation defines four internal variables. These variables are used to receive the configuration parameters from the controller (see note (1)). Note that the first action of this operation is to initialize the "random" class (see note (2)), before it goes to its idle state. Observe in note (3) that we have added some parameters to the configure message: Lose Rate from A to B (LR_AB), Lose Rate from B to A (LR_BA), Maximal Delay from A to B (MD_AB), and Maximal Delay from B to A (MD_BA).

---

17 The term "part" refers to an instance of a given class (let us say A) which composes another class (let us say B). We can say that class A is contained by class B. This instance (A) is destroyed when the container class (B) is destroyed. Take into account that this containing relationship refers to the "composition" association. For more details on parts and compositions see section 9.3.12, Property, in [SSU 10].

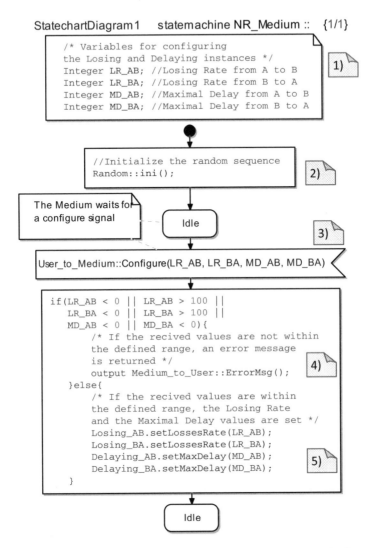

StatechartDiagram 1   statemachine NR_Medium ::   {1/1}

```
/* Variables for configuring
the Losing and Delaying instances */
Integer LR_AB; //Losing Rate from A to B          1)
Integer LR_BA; //Losing Rate from B to A
Integer MD_AB; //Maximal Delay from A to B
Integer MD_BA; //Maximal Delay from B to A
```

```
//Initialize the random sequence
Random::ini();                                    2)
```

The Medium waits for
a configure signal   Idle   3)

User_to_Medium::Configure(LR_AB, LR_BA, MD_AB, MD_BA)

```
if(LR_AB < 0 || LR_AB > 100 ||
   LR_BA < 0 || LR_BA > 100 ||
   MD_AB < 0 || MD_BA < 0){
       /* If the recived values are not within
       the defined range, an error message
       is returned */                                4)
       output Medium_to_User::ErrorMsg();
   }else{
       /* If the recived values are within
       the defined range, the Losing Rate
       and the Maximal Delay values are set */
       Losing_AB.setLossesRate(LR_AB);
       Losing_BA.setLossesRate(LR_BA);
       Delaying_AB.setMaxDelay(MD_AB);               5)
       Delaying_BA.setMaxDelay(MD_BA);
   }
```

Idle

Figure 4.32. *Non-reliable medium. State machine for an "initialize"*
operation in the NR_Medium class

On receipt of the configure message, the system validates that the received parameters are within the defined limits. If an error is found, then the ErrorMsg message is sent back to the controller (see note (4)). Otherwise, if no error is found, then the system calls all the set operations in its internal parts (see note (5)). Finally, the system goes back to its idle state.

4.6. Validation

Let us simulate the model that we have just created and compare the sequence diagrams that we obtain with those defined in section 4.4.1.

Figure 4.33 shows the lifelines[18] displayed when launching the simulation. As you can see, there are two instances of Losing class, two instances of Delaying class and one instance of NR_Medium. Also note the presence of three ports, which represent the entry points of the NR_Medium class. Finally, note that all the active classes start in their idle state, waiting for an external stimulus.

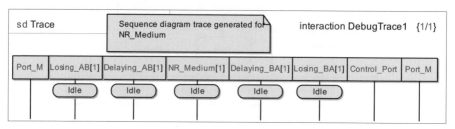

Figure 4.33. *Non-reliable medium. Initial lifelines for simulation*

4.6.1. *Configure*

We defined the expected behavior for the configuration scenario in section 4.4.1. The configuration scenario has two possible outcomes: either the received configuration parameters are correct or not.

Remember that the losing rate is defined between 0 and 100 while MAX_DELAY is defined as non-negative.

In this simulation scenario, we define the following configuration parameters: 10, 20, 30, 40 (i.e. 10% of losses from A to B, 20% of losses from B to A, 30 time units of maximal delay from A to B, and 40 time units of maximal delay from B to A).

Figure 4.34 shows the sequence diagram obtained from the simulation.

We can observe that NR_Medium receives the configure message from the Control_Port. Note that NR_Medium executes a Remote Process Call (RPC) on each one of its parts. Then NR_Medium has to wait for the end of each RPC before executing the next call.

---

18 A lifeline represents an individual participant in the Interaction. For more details on lifelines see 14.3.19, Lifeline (from Basic Interactions, Fragments) in [SSU 10].

Note also the difference between an asynchronous message (note (1)) and a synchronous message (note (2)). An asynchronous message is represented in UML by a straight arrow with an open head, while a synchronous message is represented by a straight arrow with a filled head[19]. An asynchronous message has no return signal, whilst the return signal of a synchronous message is represented by a dotted-line arrow (see, for example, note (3)).

Finally, observe that all the active classes return to their idle states at the end of the scenario.

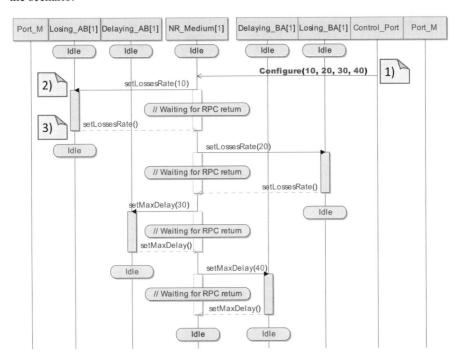

Figure 4.34. *Non-reliable medium. Simulation of the configure signal*

We can now verify the behavior of NR_Medium when an erroneous value is passed in the configure signal. This time, we use the following values: 110, 20, 30 and 40.

Figure 4.35 shows the sequence diagram obtained from the simulation. Again, all the active classes are in their idle state, waiting for a stimulus.

---

19 For more detail on all the types of messages in sequence diagrams in UML, see section 14.3.20, Message, in [SSU 10].

On receipt of the configure message from the controller, NR_Medium verifies the parameter correctness. In this scenario, the value passed for the loss rate from A to B is higher than the specified one. Therefore, all the configuration values are discarded and an ErrorMsg message is sent back to the controller user.

Note that no further explanation is given to the controller user. It is obvious that it would have been possible to provide an explanation within the ErrorMsg message; however, we decided to define a message that was as simple as possible.

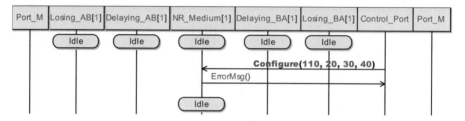

Figure 4.35. *Non-reliable medium. Error message for the configure signal*

You can compare the sequence diagrams obtained from this simulation against those defining the expected behavior described in Figure 4.20 in section 4.4.1, and verify that the simulation correctly matches the expected behavior.

4.6.2. Send

In order to test the send scenario, we can define four different cases which will allow us to test the losing and delaying behaviors separately. These four cases are:

– neither losses nor delay;

– losing everything;

– no losses and high delay; and

– losses and delay.

4.6.2.1. *Neither losses nor delay*

In this scenario, we configure the medium for neither losing nor delaying the data packets. We should observe that all the messages are instantly delivered to the corresponding receiver.

We can configure the medium with the following parameters: 0, 0, 0 and 0 (0% of losses from A to B, 0% of losses from B to A, 0 time units of maximal delay from A to B, and 0 time units of maximal delay from B to A).

Figure 4.36 shows the sequence diagram obtained when sending two consecutive data messages from A to B.

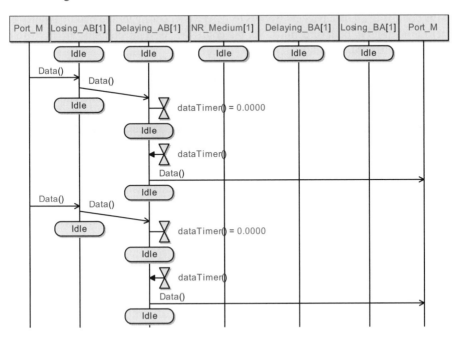

Figure 4.36. *Non-reliable medium. Sending data with neither losses nor delay from A to B*

Figure 4.37 shows the sequence diagram obtained when sending a data message from B to A. As we could have expected, the obtained behavior is symmetrical with the previous diagram.

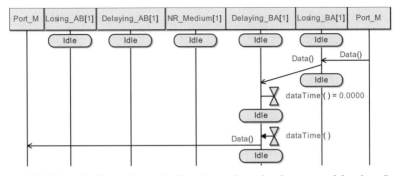

Figure 4.37. *Non-reliable medium. Sending data with neither losses nor delay from B to A*

4.6.2.2. *Losing everything*

In this scenario, we configure the medium for not losing the data packets from A to B but losing all data packets from B to A. We also configure the maximal delay to zero time-units in both directions. So, the configure message contains the following parameters: 0, 100, 0, and 0.

In Figure 4.38 you see the sequence diagram obtained from simulation. As you will observe in notes 1 and 2, a data packet was sent from A to B and another in the opposite direction. Note that the first message was successfully delivered to its destination while the second was lost. We have sent a second set of data messages with the same results.

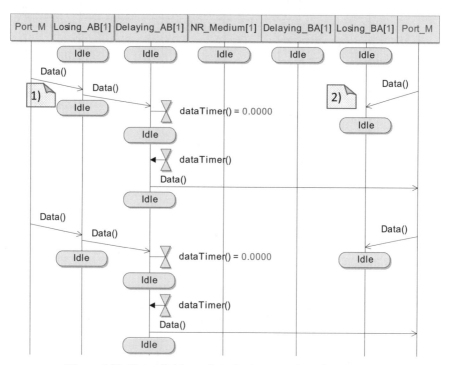

Figure 4.38. *Non-reliable medium. Losing everything from B to A*

We can repeat the test many times and we will always obtain the same result (see Figure 4.39).

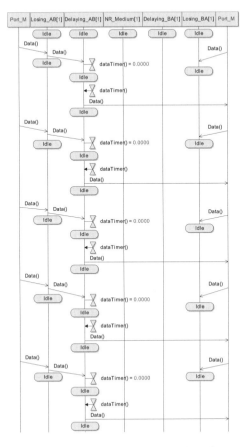

Figure 4.39. *Non-reliable medium. Losing everything from B to A (cont.)*

You can compare the obtained result against the expected behavior defined in section 4.4.1 and verify that our results correctly match the expected behavior.

4.6.2.3. *No losses and high delay*

In this scenario, we will configure the medium for not losing the data packets in any direction. We will also configure the maximal delay to 100 time-units from A to B and to 500 time-units from B to A.

Figure 4.40 gives the sequence diagram obtained from the simulation. You will observe that no message was lost in the simulation. Note that the data packet sent by B (note (1)) has a delay of 10.0406 time-units, while the data packet sent by A has a delay of 58.9862 time-units. Observe that, as expected, the data packet going from B to A is delivered before the data packet going from A to B.

This simulation allows us to give some details about time in our simulation scenarios.

It is important to note that we have talked about time-units in a general way. Indeed, as might be expected, our simulations do not take into account real time, but rather a simulated time. As a consequence, all we can guarantee is that when we define a timer A as being lower than a timer B, then timer A will be triggered before timer B.

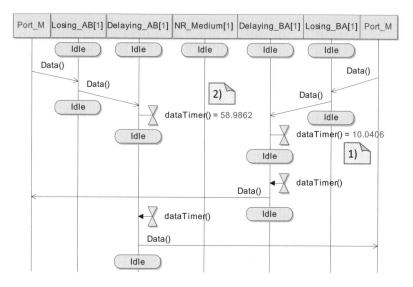

Figure 4.40. *Non-reliable medium. Random delay from B to A*

Figure 4.41 shows a second execution of the same simulation. In this simulation, the packet going from A to B has a delay of 56.9628 time-units, while a packet in the opposite direction has a delay of 452.6353 time-units.

Again, we can observe that the time sequence is correctly respected. In this case, the data packet going from A to B is delivered before the data packet going in the opposite direction.

Note here another detail about the simulated time used in our experiments: the data packet going from B to A experiences a delay almost ten times higher than the packets in the opposite direction. However, our simulation does not show any sign of this ratio since the packets are placed just one after the other in the sequence diagram. Nevertheless, the smaller timer is triggered before the other one.

A great advantage of this simulated time is that we can execute simulations containing very long delays without being subjected to the limitations of a real-time duration. The inconvenience of such simulated time is that it is not possible to deduce any time performance measurements from it.

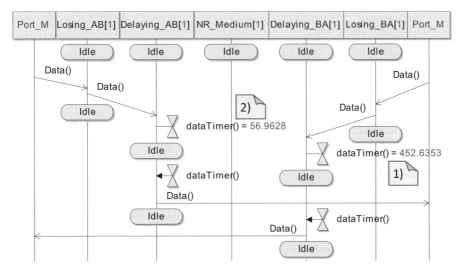

Figure 4.41. *Non-reliable medium. Random delay from B to A*

4.6.2.4. *Losses and delay*

In this scenario we configure the medium in order to have a completely reliable but very slow communication from A to B. We also configure the medium in order to have a very fast but very unreliable communication from B to A. This kind of configuration could correspond to a TCP communication from A to B and a UDP communication from B to A.

The configuration that we use for this simulation is the following: no losses from A to B, 80% of losses from B to A, a delay of 500 time-units from A to B and 10 time-units from B to A.

Figure 4.42 shows part of our simulation. You can observe that we sent one data packet from A to B (see note (1)) and another from B to A (see note (2)). Note also that the message from B to A is discarded by the Losing_BA module (it is not sent to Delaying_BA). Before the timeout related to the data packet sent by A to B, we try to send another message from B to A (see note (3)), but the packet is also discarded by Losing_BA. Finally, after the timeout interruption (see note (4)), the data packet is successfully delivered to B.

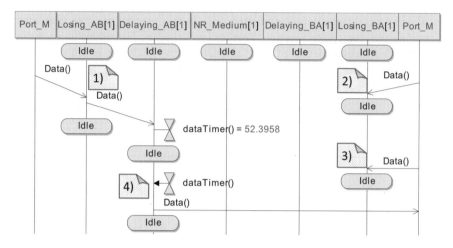

Figure 4.42. *Non-reliable medium. Opposite performances in each direction (1)*

Let us execute this simulation again. Figure 4.43 shows the second execution of this scenario. Again, we have sent a data packet in both directions simultaneously. As you can see, this time the packet going from B to A is not lost by the Losing_BA module. After that, you can see that the data packet sent from A to B is very delayed; then, it is delivered to its destination after the other one.

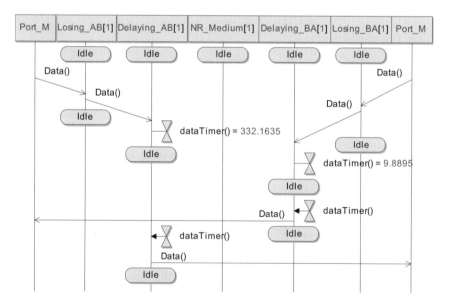

Figure 4.43. *Non-reliable medium. Opposite performances in each direction (2)*

Executing the scenario yet again, we obtain a very interesting result. Figure 4.44 shows this new sequence diagram.

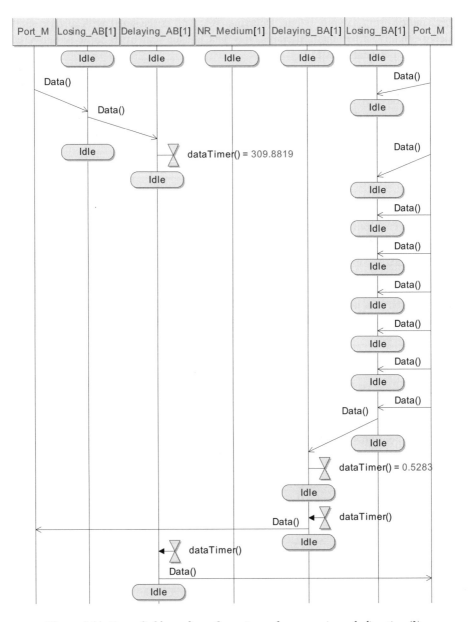

Figure 4.44. *Non-reliable medium. Opposite performances in each direction (3)*

In this new diagram you can firstly observe that the data packet going from A to B was assigned with a very high delay. After that, you can observe that many successive messages are lost from B to A. After many attempts, a data packet is transmitted from Losing_BA to Delaying_BA, and is assigned with a very short delay. Finally, after many attempts, the data packet sent by B to A is delivered before the one sent by A to B.

This simulation perfectly illustrates the benefits and drawbacks of two of the best known protocols, TCP and UDP. When a network is very congested, a heavy protocol such as TCP might be useless for transmitting real-time data because the information might be outdated when the destination application receives it.

On the other hand, a very light protocol such as UDP might be useless for applications requiring full reliability.

This simulation also allows us to imagine many different solutions between UDP and TCP, each one adapted to a different data flow, and we can see how a module such as "Non-Reliable medium" could be useful in order to test and simulate such new protocols.

However, a deeper investigation of this kind of protocol is not within the scope of this book and here we only give a few comments which might motivate further study.

4.6.3. *Extending the model*

4.6.3.1. *Adding a payload*

Generally, a medium should be able to manage any kind of data type; nevertheless, implementing this capability in our graphical model would be too long and this type of detail is also beyond the scope of this book.

Instead, here we only add a simple integer parameter in order to illustrate how the model could work under these circumstances. Of course, the implementation of a general solution might be more or less easy to implement, depending on the development language used.

First, we modify the list of messages defined in our interfaces (see Figure 4.15 in section 4.4). Figure 4.45 shows how the data packet must be modified in the interfaces.

ClaDiag_NRMedium_Interfaces_03

<<interface>> **App_to_Medium**	<<interface>> **Medium_to_App**	<<interface>> **Los_to_Del**
+Data(Payload: int)	+Data(Payload: int)	+Data(Payload: int)

Figure 4.45. *Non-reliable medium. Interfaces, data packets with parameters*

Now we need to update the state machine defining the behavior of the losing module previously described in Figure 4.27 in section 4.5.2.2.

Figure 4.46 gives the modifications performed. First, in note (1) you can see that we added a new integer attribute called payload. In note (2), you can see that we now receive a data packet containing a parameter. This parameter is stored in the new payload attribute. Finally, in note (3), we send a data message containing the previously set payload attribute.

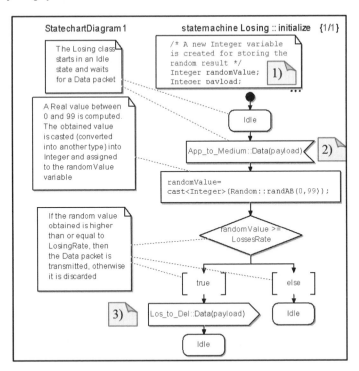

Figure 4.46. *Non-reliable medium. Using the payload attribute in the losing class*

We also need to update the state machine previously defined for the delaying class in Figures 4.30 and 4.31 in section 4.5.3.2.

In Figure 4.47 you can see the modifications performed. In note (1) you can see a new integer parameter called payload. This parameter is used in order to store the received data (see note (2)). Finally, this parameter is sent to the receiver application (see note (3)).

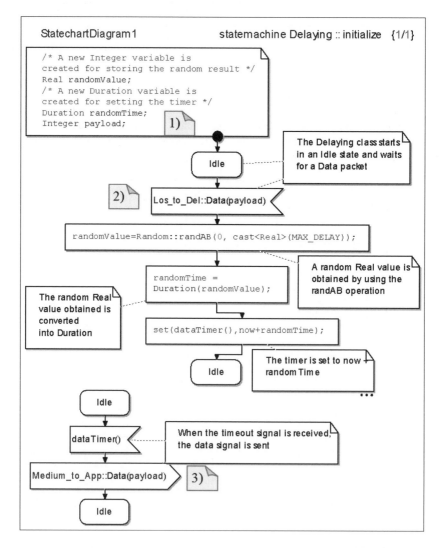

Figure 4.47. *Non-reliable medium. Using the payload attribute in the delaying class*

In order to test the new state machines, we repeat the simulations performed in section 4.6.2.4. For this, we configure the medium as follows: no losses from A to B, 80% of losses from B to A, a delay of 500 time-units from A to B, and 10 time-units from B to A.

As you can see in Figure 4.48, this diagram correctly matches those obtained in Figures 4.42, 4.43 and 4.44 in section 4.6.2.4. The only difference is that this time we see the payload being transmitted within the data packet.

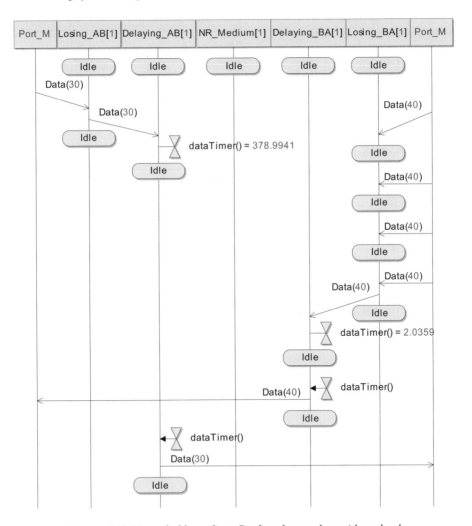

Figure 4.48. *Non-reliable medium. Sending data packets with payload*

4.6.3.2. *Consecutive data packets*

At this point, we have verified the losing and delaying modules separately. We have also tested how our system reacts when receiving two simultaneous messages from different entities; and we have verified that there is no interference between them since each flow has its own set of controller modules.

Let us now verify what happens if the medium receives two consecutive data packets through the same port. For this test, we need to configure the medium as being completely reliable and quite fast. The parameters to be used by our simulation are 0, 0, 10 and 10.

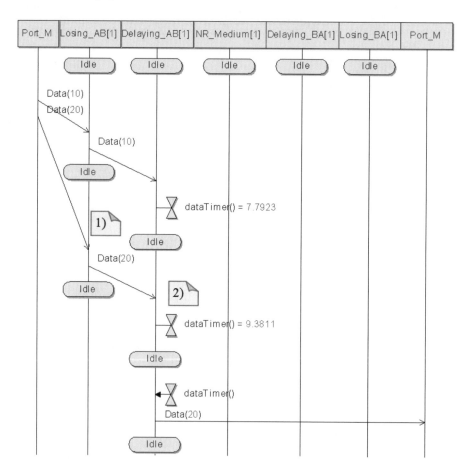

Figure 4.49. *Non-reliable medium. Sending two consecutive data packets to a single port*

As you can see in Figure 4.49, we sent two consecutive data packets, one containing a value 10 and the other containing a value 20. You can see that the losing class correctly processes the data packets since the time required to do this is negligible (see note (1)).

By contrast, the delaying class receives the second data packet before the timeout of the first packet (see note (2)). Under this condition, the delaying class sets the timer with the new time before the expiration of the previous set operation. The consequence is that the delaying class discards the first data packet and only the second packet is delivered to its destination.

We can solve this problem by taking advantage of the timer management in our simulation system. A timer can have an identifier. Hence, every time we set a timer with a different identifier, we are creating a new instance of a timer.

Thus, we can use this identifier in order to carry the received payload. In this way, when the delaying class receives a timeout, it takes the timer identifier and sends it in a data packet to the receiver entity.

Now, we need to modify the state chart diagram, setting the timer within the delaying class. Figure 4.50 shows the new timer with an identifier carrying the payload.

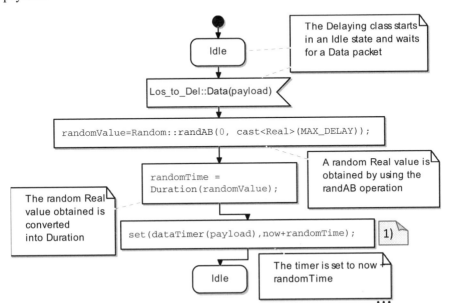

Figure 4.50. *Non-reliable medium. Adding a payload to the timer*

Finally, we need to read the parameter in the timer at timeout. Figure 4.51 shows the state chart reading the payload from the timer and sending it to the receiver entity.

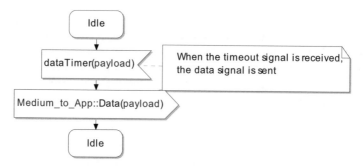

Figure 4.51. *Non-reliable medium. Reading the payload from the timer*

Let us repeat the simulation in order to validate our solution.

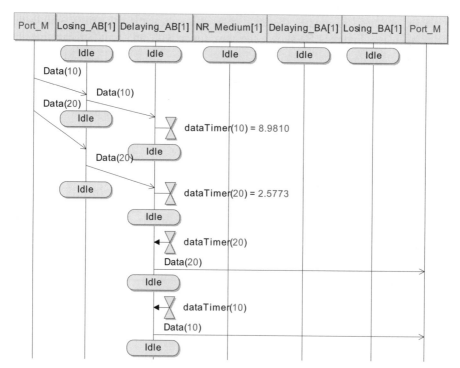

Figure 4.52. *Sending two data packets simultaneously*

In Figure 4.52 we see that no data packet is lost this time. We can also note that the delay assigned to data (10) is higher than the one assigned to data (20). Finally, we observe that the medium can manage this kind of situation, i.e. data (20) is delivered before data (10).

We have taken advantage of the simulation software features in order to manage multiple timers simultaneously. Nevertheless, as you can imagine, there may exist many different solutions to this multiple timers problem, for example a parallel state machine or a multithreading process. However, we have preferred to present a very simple solution that illustrates that many functional omissions, or gaps and design bugs, can be discovered and solved using only a detailed model and some simulations.

As a bonus example, consider Figure 4.53. In this simulation, we have sent several consecutive and simultaneous data packets to both ports.

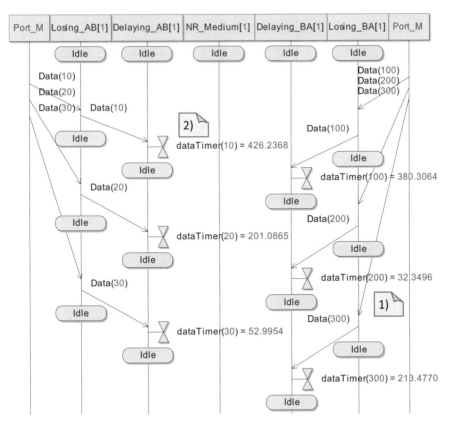

Figure 4.53. *Non-reliable medium. Sending consecutive and simultaneous data packets*

As you can see, on one side we have sent three data packets containing the following values: 10, 20 and 30. On the other side, we have sent three data packets containing the following values: 100, 200 and 300.

Observe the timer value assigned to each data packet. Note that the smaller timer is the one assigned to Data(200) (see note (1)). Also note that the higher timer is the one assigned to Data(10) (see note (2)). If we organize the data packets in an increasing order with reference to their timers, we obtain the following list: Data(200), Data(30), Data(20), Data(300), Data(100), Data(10). If our model works properly, then no packet will be lost and the order of the timers will be preserved.

Figure 4.54 shows a continuation of this simulation.

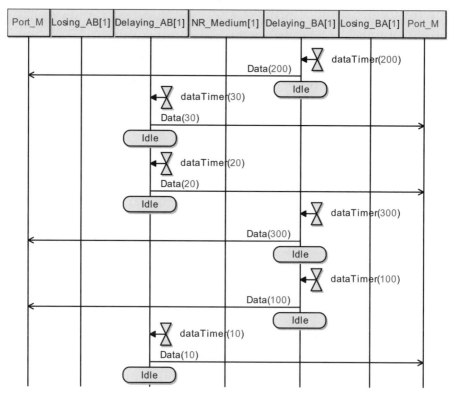

Figure 4.54. *Non-reliable medium. Sending consecutive and simultaneous data packets (continued)*

This simulation confirms that no packet is lost due to a model error. Moreover, you can observe that the timers properly correspond to our expectations.

We can therefore consider that our model correctly corresponds to the requirements defined at the beginning of this chapter.

Note that this last simulation leads to another interesting problem within the network framework: a disordered data flow. We do not propose a solution to this problem here, as our goal is only to show how these network problems can be modeled and simulated through a UML model, in order to help readers to model their own problems and solutions.

4.7. Chapter summary

In this chapter we have shown how to model a non-reliable and configurable communication medium. Such a model is very useful in order to validate higher level protocols and mechanisms.

We have also shown how to improve a model by adding some code. We have illustrated once more how a detailed model can help a designer to discover possible gaps or ambiguities in the requirement specification. We have seen that a simulation can highlight some deficiencies in the design phase. Finally, we have shown that these simulations can help us to correct and validate the initial design and, in particular, the modifications performed in this initial design which occur during the validation phase.

4.8. Bibliography

[RFC 68] RFC 768, User Datagram Protocol, Specification IETF, Defense Advanced Research Projects Agency, http://tools.ietf.org/html/rfc768

[RFC 93] RFC 793, Transmission Control Protocol, Specification IETF, Defense Advanced Research Projects Agency, http://tools.ietf.org/html/rfc793.

[SSU 10] Unified Modeling Language™ (UML®), Superstructure specification, 2010-05-05, http://www.omg.org/spec/UML/2.3.

Chapter 5

Simple Transport Protocol

5.1. Introduction

Transport protocols hide all the possible problems inherent to the physical and network layers, such as disorder, losses, delay and corruption. These protocols provide a reliable and ordered communication between a sender and a receiver, the usual strategy being to retransmit a lost or corrupted packet. At the same time, the most sophisticated of these protocols implement different mechanisms in order to avoid duplication and ensure correct delivery order, and to optimize the performance of data delivery versus the underlying real or virtual network.

In this section, we will create our own, simple transport protocol to provide a correct order and reliability to the user layers when transmitting a set of messages over a Non-Reliable Medium. To do this, we will use the well-known basic algorithm called the Alternating Bit Protocol.

To reach this goal, this chapter builds a system architecture by linking the Simple Chat Application modeled in Chapter 3 to the Non-Reliable Medium modeled in Chapter 4, through a Simple Transport Protocol.

Of course, to do so, it will be necessary to slightly adapt and modify the precedent models, and to extend the characteristics of the transmitted data type.

5.2. Requirements

Two remote chat applications need to communicate through a Non-Reliable Medium, and we therefore propose a transport protocol with the following characteristics:

– the protocol will offer a reliable and ordered communication service to its application;

– the protocol will support a full duplex (bidirectional) communication;

– the protocol will ensure the correct delivery of a data packet before accepting a new data packet to transmit, i.e. it accepts and transmits one single packet at a time from the application;

– the protocol instances communicate through a Non-Reliable Medium;

– the protocol will implement the Alternating Bit mechanism in order to solve packets losses and delays;

– the communication interfaces of the application and the medium are already defined because the modules already exist, and the transport protocol will use the interfaces defined in those modules.

5.3. The Alternating Bit Protocol

The Alternating Bit Protocol (ABP) is a reliable transport protocol solving the problem of transferring data from a sender A to a receiver B through a Non-Reliable Medium. As it is simple and very illustrative (although not efficient), we will use it in order to show how to design such a transport protocol and how to link it to its adjacent layers.

First we will describe the Alternating Bit Protocol by using the Petri Nets formalism, since this is one of the traditional representations of this algorithm in the network context. After that, we will explain how to model this algorithm in UML.

But before presenting the algorithm, let us recall some of the Petri Net notions needed for correct understanding of the ABP's behavior and of the Petri Net model deduced.

5.3.1. *Basic communication features with Petri Nets*

We can represent the sending and receiving process actions by using Petri Nets, as shown in Figure 5.1.

In this figure, the transition labeled !E represents the execution of the sending process of data E, while the transition labeled ?E represents the execution of the receiving process of data E. As you can see, in this model, there is no explicit link between the sender and the receiver, i.e. the underlying sending-receiving interaction is not represented.

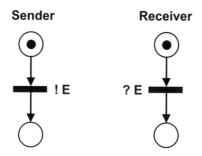

Figure 5.1. *Simple transport protocol. Alternating bit protocol, representing data sent and received in Petri Nets*

We can also represent the same behavior with an explicit representation of the sending-receiving interaction by adding a place E between the sender and the receiver, as shown in Figure 5.2. A mark or a "token", as explained in section 1.1.2.2.2, in this new place represents that the data was sent by the sender, and that it has not yet been received; in other words, it represents the fact that the message is in transit in the communication medium (or the network).

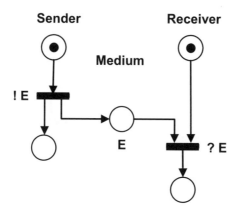

Figure 5.2. *Simple transport protocol. Alternating bit protocol: sending and receiving with a shared place*

Figure 5.2 represents a shared-place communication. Indeed, a token in this place allows the firing of transition ?E. When !E is fired, the token is removed from place Sender, and one token is added to each of the two places that are the outputs of the transition !E, including place E. A token in place E means that the message is in transit in the medium. Transition ?E can now be fired since there is a token on each input place in this transition. Firing this transition means that the message is being received. After firing ?E, the two tokens are removed from the places Receiver and E, and one token is added in the output place of transition ?E. This means that the receiver has received the message and that is going to process it.

By using the same mechanism, we can represent a very simple protocol defining an entity sending a data packet and waiting for an acknowledgement answer, as shown in Figure 5.3.

Observe that, on the right side, the Receiver sends an acknowledgement ackE message. This action is represented by the transition labeled !ackE. On the left, the reception of the ackE message is represented by the transition labeled ?ackE. In the middle, a token in the place labeled ackE represents that the acknowledgement message was sent but is still not received; in other words, it is in transit through the communication medium.

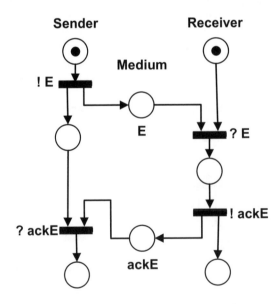

Figure 5.3. *Simple transport protocol. Alternating bit protocol:
sending data and waiting for acknowledgement*

In a real communication network, the data packets may be lost.

Figure 5.4 shows the Petri Net of Figure 5.2, extended to describe the data losses.

Let us observe the new transition, labeled "Loss". The firing of this transition, which is possible when there is a token in place E (i.e. a message in transit), represents the medium losing the data stored in place E, as the token is removed from place E and will disappear.

Note that there are two possible behaviors when place E is marked: either transition "Loss" is fired and the message is lost, or transition ?E is fired, and the message is received. These are the two possible behaviors that our simple protocol has to handle when messages can be lost.

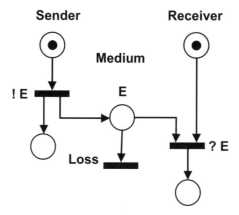

Figure 5.4. *Simple transport protocol. Alternating bit protocol: representing a loss*

In this model, if the data packet E is lost, then the ?E transition cannot be fired by the receiver and the system is blocked. Under these circumstances, the sender needs to resend the data packet. Figure 5.5 shows a new model where the firing of the "Loss" transition brings a mark back to the initial place on the sender's side. This mark allows the sender to send a data packet again. Note that this process is repeated as long as the data packet is lost.

Note that the acknowledge packet might be lost by the medium in the same manner as the data packet. On the sender side, the only observable event is a missing acknowledgement packet. The problem here is that the sender cannot know if the lost message was the E data packet or the acknowledgement. However, in both circumstances, the sender needs to resend the data packet.

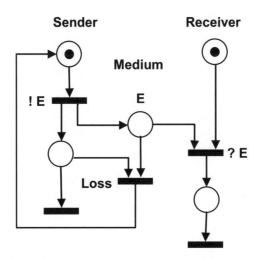

Figure 5.5. *Simple transport protocol. Alternating bit protocol: resending data after a loss*

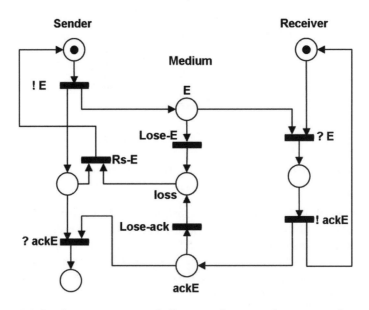

Figure 5.6. *Simple transport protocol. Alternating bit protocol: recovering from a loss*

Figure 5.6 represents the loss of the acknowledgement as a new transition labeled "Lose-ack". At the same time, we have renamed the transition representing the loss of the E data packet as "Lose-E". Finally, we have represented the loss of either of these two messages as a new place, labeled "loss". Note that this new place can receive a token from either Lose-E or Lose-ack transitions. A token in this loss place, while the sender is in the middle left state (message sent but not yet acknowledged), allows the firing of the transition Rs-E, and this transition leads to the sender place and to the retransmission of the data packet.

This model seems to work correctly. However, it presents a problem when the ackE packet is lost. Consider the following scenario:

– the sender sends a data packet (transition !E is fired);

– the data is received on the receiver side (transition ?E is fired);

– the receiver sends an ackE packet (transition !ackE is fired);

– the acknowledgement packet is lost (transition Lose-ack is fired);

– the sender is allowed to resend data (transition Rs-E is fired);

– the sender sends a data packet (transition !E is fired);

– the data is received on the receiver side (transition ?E is fired);

– the receiver sends an ackE packet (transition !ackE is fired).

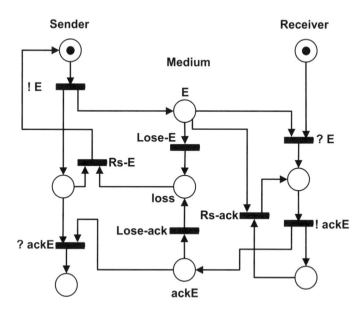

Figure 5.7. *Simple transport protocol. Alternating bit protocol: avoiding duplicated data*

At this point, the data packet E is received twice by the receiver. We need a mechanism to detect that the data packet was already received by the receiver.

Figure 5.7 shows a new model with a solution to the previous problem. Observe that there is a new transition labeled "Rs-ack". Note that this transition is enabled only when a data packet is in transit through the network and an acknowledgement packet was already sent by the receiver.

Observe that the firing of this transition has to be defined in such a way that the receiver must ignore the duplicated packet and retransmit the acknowledgement packet by firing the transition !ackE again.

5.3.2. *The algorithm*

The algorithm of the Alternating Bit Protocol (ABP) has to provide a reliable transport protocol solving the problem of transferring data in a unidirectional mode from a sender A to a receiver B through a Non-Reliable Medium. Note that it is a simplified form of the well-known Sliding Window mechanism with a window size of 1.

The ABP algorithm creates data packets containing the payload to be transferred. It then associates a control bit with each data packet, and the control bit is used to ensure a correct delivery order simply alternates between 0 and 1.

It also uses positive acknowledgements (ACK) in order to confirm the reception of data. Indeed, only one bit is needed since no new data is sent before acknowledgement of the previous data packet is received by the sender (stop-and-wait).

To begin, sender A sends a data packet (payload + control bit) and continues sending the same data at a regular frequency. Sender A stops sending the data packet when it receives an acknowledgement containing the same control bit from receiver B. After that, sender A complements (flips) the control bit and starts sending the next data packet containing the new control bit.

On the receiver side, on receipt of a data packet from sender A, receiver B sends back an acknowledgement packet containing the control bit that is included in the received message.

The first time B receives a message with a given control bit, it sends the payload to its receiver client for processing. If subsequent data packets containing the same control bit are received, they are only acknowledged (without delivering the message to the client).

Figure 5.8 shows a representation of the ABP [DIA 09]. Note that this model is an extension of the one presented in Figure 5.7.

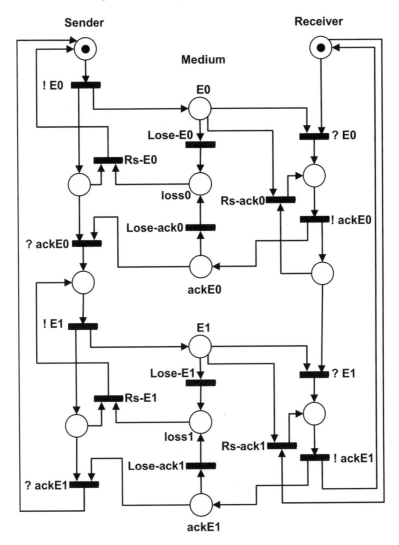

Figure 5.8. *Simple transport protocol. Algorithm for the ABP*

In this new model, we represent a data packet containing a payload and a control bit as two transitions labeled E0 (!E0 for sending and ?E0 for receiving) for the data packet containing the control bit 0, and E1 (!E1 for sending and ?E1 for receiving) for the data packet containing the control bit 1. Note that the transitions labeled ?E0 and ?E1 represent the receiver reading the payload from a data packet and sending it to its client for processing. Note also that the sender does not send any further message before receipt of an acknowledgement for the currently sent data packet. Finally, observe that the sender and the receiver come back to their initial state after receipt of ackE1.

We can represent the same algorithm by using two separate state machines, one for the sender and one for the receiver. For simplicity's sake, we will represent the exchanged information as only E(ControlBit) and Ack(ControlBit). We will add the details concerning the payload in a later section when defining the protocol behavior.

Figure 5.9 shows a simple state machine representing the sender. Note here that the sender resends the data only when its internal timer is over or when it receives an unexpected message from the receiver (states St_WaitingAck_0 and St_WaitingAck_1). After receiving the expected Ack message, the sender waits for a new data packet from its sender application (states St_WaitingData-0 and St_WaitingData-1). Any Ack is ignored while waiting for a new data.

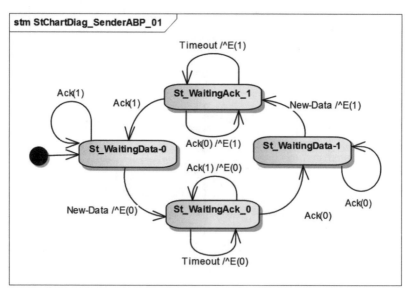

Figure 5.9. *Simple transport protocol. Sender's behavior for the ABP*

Figure 5.10 shows a simple state machine representing the receiver behavior. Observe that the receiver only processes the received data the first time; after that, it acknowledges the data packet but does not process it.

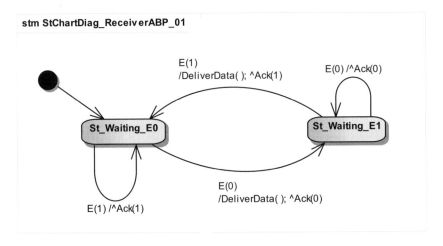

Figure 5.10. *Simple transport protocol. Receiver's behavior for ABP*

5.4. Analysis

We will follow an iterative and incremental approach in order to analyze our system. This strategy will allow us to find the expected interfaces, their messages and some accessory classes.

5.4.1. *First step: The Alternating Bit Protocol*

We will now analyze the messages sent and received by the ABP without taking into account the application and the Non-Reliable Medium. The simplest way to do this is to consider the Alternating Bit Protocol Sender (ABPS) and the Alternating Bit Protocol Receiver (ABPR) as two separate systems. By doing so, we obtain two sets of messages.

Figure 5.11 shows the messages sent and received by the ABPS. Note that this diagram represents the sending process of a single data packet and does not take into account the change of the Control Bit (CB). Observe that this diagram corresponds to the Petri Nets model described in section 5.3.2.

User_A sends a data packet to the ABPS. Then, the ABPS sets its internal timer (message 2) and sends an E message containing a control bit and the received data to the ABPR (message 3). While no correct acknowledgement is received after a given time (timeout in message 4, and erroneous CB in message 6), the ABPS sends the E message again (messages 5 and 7). Finally, on the receipt of an acknowledgement message containing the expected CB (message 8), the ABPS resets its timer (message 9) and waits for a new data from the user.

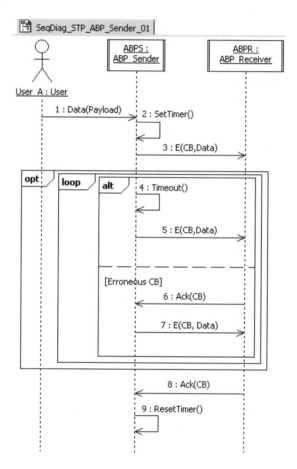

Figure 5.11. *Simple transport protocol. Sequence diagram for an Alternating Bit Protocol Sender (ABPS)*

Figure 5.12 shows the messages sent and received by the ABPR. On receipt of an E packet, if the received data was not previously received, then the ABPR delivers the received data to its user; otherwise, the data is ignored. In any case, the received message is acknowledged with the received CB.

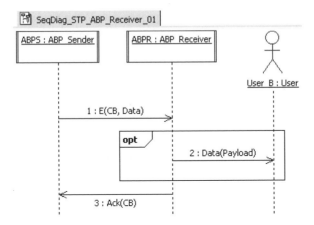

Figure 5.12. *Simple transport protocol. Sequence diagram for an Alternating Bit Protocol Receiver (ABPR)*

These sequence diagrams lead us to believe that the set of exchanged messages is as shown in Figure 5.13.

In this diagram we can see that the ABPS receives Data(Payload) from the user. It also sends E(CB, Payload) and receives Ack(CB) to/from the NRM.

The ABPR sends Data(Payload) to the user. It also sends Ack(CB) and receives E(CB, Payload) from the NRM.

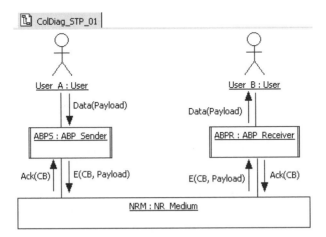

Figure 5.13. *Simple transport protocol. Messages exchanged between the ABP and its actors*

However, it is not a (virtual) user who sends data to the protocol, but an already implemented application, here a chat application. This application has a well defined interface and our protocol must use it.

5.4.2. *Second step: Linking the application*

Figure 5.14 recalls the messages sent and read by the current chat application model (defined in section 3.3.3). Note that there are eight different messages exchanged between the two instances of the chat application.

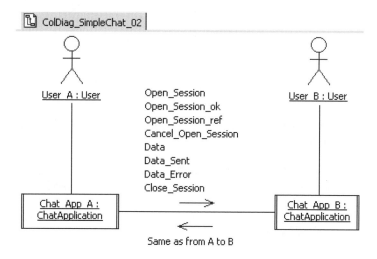

Figure 5.14. *Simple chat. List of messages between chat application instances*

We can modify the data packet in order to add a payload. Let us say that the payload carried by this packet is a character string. By doing so, the new data packet will be defined as Data(Charstring: Payload). The rest of the messages will remain as they are.

Observe that the messages received by the ABPS and those sent by the ABPR are much more complex than simply Data(payload).

Currently, the ABPS only reads Data(payload) from the application and sends E(CB, Payload) to the ABPR. If we want the ABPS to read all the possible messages from the application, then we will need to create and manage many messages from ABPS to ABPR, i.e. E(CB, Open_Session), E(CB, Close_Session), E(CB, Data(Payload)), etc.

However, remember that our protocol must be indifferent to the transported data. In order to avoid a complicated algorithm, we need to define an encapsulating packet able to represent any of the application packets, in an homogeneous and general way.

To do so, we can create an Application Data Unit (ADU). An ADU contains a description of the represented message and of the payload carried. The structure of an ADU is as follows: ADU(Header, Payload). This way, a message Data(Payload) coming from the application corresponds to ADU("Data", Payload); the Open_Session message corresponds to an ADU("Open_Session", Empty_payload), etc.

We can also create a new module named ADU_Manager (ADUM), as we do not want to interrupt the ABPS and ABPR modules with more responsibilities and actions than necessary. The new specialized module which we create will translate the messages coming from the application into Application Data Units and vice versa.

Figure 5.15 shows a new collaboration diagram depicting the messages exchanged between the ABP modules and the new ADUM. Note that the ABPS now receives a Data(ADU) message from the ADUM while the ABPR sends the same message to the ADUM. We should also note here that the old E(CB, Payload) message becomes E(CB, ADU). Thus, the behavior of the ABP modules remain unchanged.

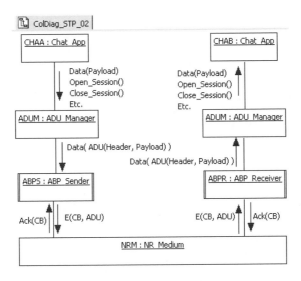

Figure 5.15. *Simple transport protocol. Messages exchanged between ABP and ADUM*

5.4.3. *Third step: Linking the medium*

Now, we can link the protocol model to the Non-Reliable Medium (NRM). The NRM has a well defined interface, described in section 4.3.3. Figure 5.16 recalls the messages sent and received by the NRM.

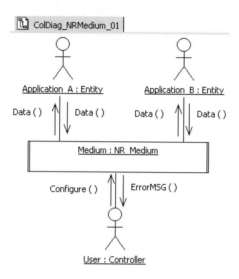

Figure 5.16. *Non-reliable medium. Messages exchanged between the medium and its actors*

Observe that the current NRM model receives and sends a unique message to the entities: the Data() packet. Moreover, remember that the NRM does not read or process the received data, as it only transfers (or not) the received information from one port to another.

Conversely, the ABP modules send two kinds of messages to the medium, E(CB, ADU) and Ack(CB).

If we leave the ABPS and ABPR modules as they are now, then we will need to update the NRM model in order to add more information to the data packet. In addition, we will need to create a different data packet for every packet coming from the ABP modules.

We can simplify and homogenize the communication between the ABP modules and the medium by changing the current medium as little as possible.

First, we can modify the data packet in order to add a payload to it. After that, we can create a new data type called a Protocol Data Unit (PDU). A PDU encapsulates the packets sent and received by the medium. The structure of a PDU is defined as follows: PDU(Header, CB, ADU). In this way, the packet E(CB, ADU) corresponds to PDU("E", CB, ADU), while the Ack(CB) packet corresponds to PDU("Ack", CB, Empty_ADU). From this, the packets sent to the medium will be as simple as Data(PDU). In this way, with minimal modification, we can encapsulate all the information exchanged between the ABP modules.

We do not want to complicate the modules implementing the ABP algorithm by adding encapsulating behavior. Instead, we can create a new module named PDU_Manager (PDUM). This new module translates the messages exchanged between the ABPS and the ABPR into PDUs and sends them to the medium. In the reverse direction, the PDUM translates the PDU packets coming from the medium into E and Ack messages, and sends them to its corresponding algorithm module.

Figure 5.17 shows the new list of messages exchanged including the new PDUM module.

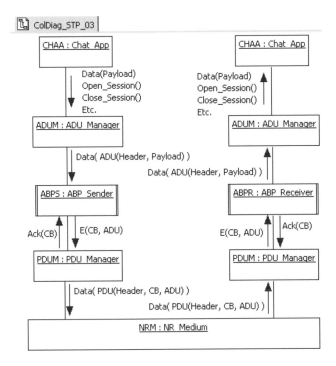

Figure 5.17. *Simple transport protocol. Messages exchanged between PDUM and NRM*

5.4.4. *Concerned classes*

Our current model is composed of four different classes, ADU_Manager, PDU_Manager, ABP_Sender and ABP_Receiver.

We can add one more class and call it Simple_Transport_Protocol (STP). This new class is a container encapsulating all the transport protocol behavior and architecture.

The class diagram describing the relationship between these classes is shown in Figure 5.18. Note that all the classes are related to the container through a composition association.

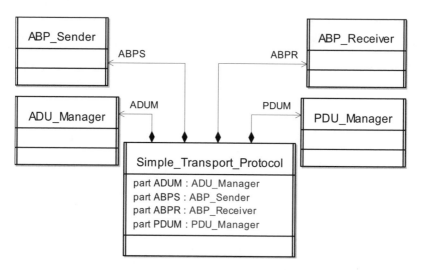

Figure 5.18. *Simple transport protocol. Class diagram representing the Simple Transport Protocol*

We also need to define two data types representing the encapsulated data used in our model.

Figure 5.19 shows these two data types. Observe that PDU is related to ADU by a composition association.

Note that the payload in ADU and the header in both classes are defined as a character string. This is not the most efficient solution, but it will allow us to better read the simulation results.

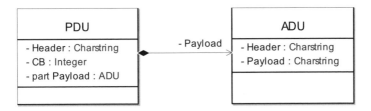

Figure 5.19. *Simple transport protocol. Encapsulated data
used by the simple transport protocol*

5.4.5. *Signal list definition*

The ABPS sends a packet E(CB, ADU) towards the ABPR. We can gather this packet into an interface called I_ABPS_ABPR. Figure 5.20 shows a detailed description of this interface. Remember that this message is translated into a PDU by the PDUM.

The ABPR sends a packet Ack(CB) towards the ABPS. We gather this packet into an interface called I_ABPR_ABPS. This message is translated into a PDU by the PDUM.

The ABPS receives a Data(ADU) packet from the application. Remember that the ADUM intercepts the messages coming from the application and translates them into ADUs. We can gather this packet into an interface called I_App_ABPS.

Finally, the ABPR sends a Data(ADU) packet towards the application. This packet is translated into application-styled packets by the ADUM. We will gather this packet into an interface called I_ABPR_App.

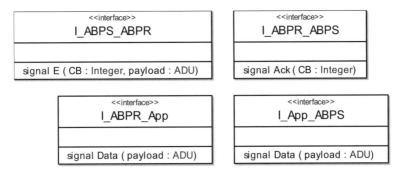

Figure 5.20. *Simple transport protocol. Interfaces
used by the simple transport protocol*

We will use the interfaces defined previously in section 4.3.3 to communicate with the NRM. We can modify the Data packet in order to add a payload. The new interface list is shown in Figure 5.21.

Note that we changed the name of the data packet to Data_AtM and Data_MtA. This modification allows us to facilitate the reading of the simulation results by differentiating this packet from the other data packets.

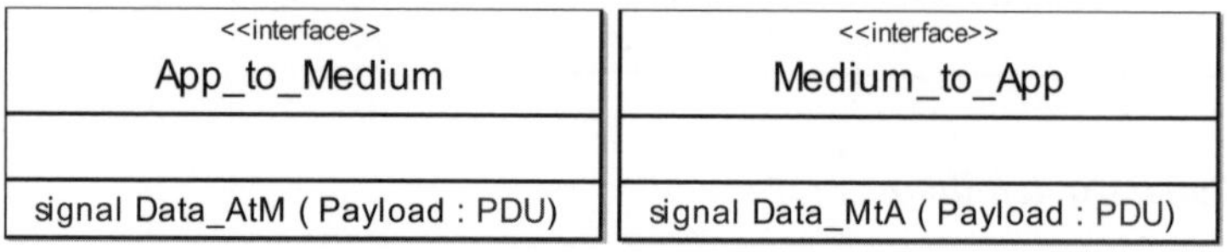

<<interface>>
App_to_Medium
signal Data_AtM (Payload : PDU)

<<interface>>
Medium_to_App
signal Data_MtA (Payload : PDU)

Figure 5.21. *Simple transport protocol. Modified interfaces for a non-reliable medium*

Finally, we can use the interface defined in section 3.3.3 to communicate with the chat applications. We can modify the data packet in order to add a payload.

Note that, additionally, we have modified the name of the data packet to DataAp (Figure 5.22). This modification facilitates the reading of the resulting simulations by differentiating this data packet from those sent by other modules.

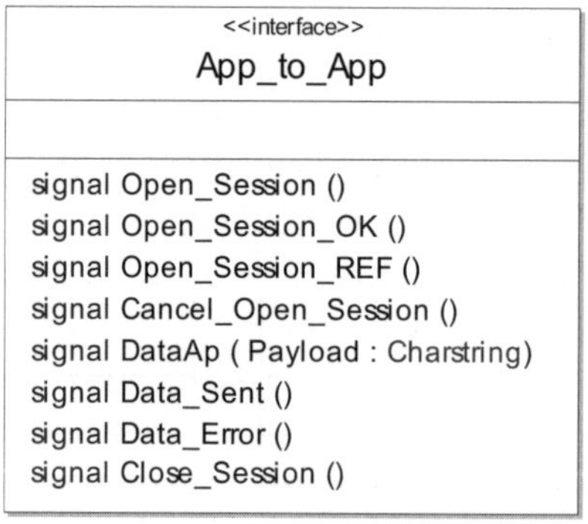

<<interface>>
App_to_App
signal Open_Session ()
signal Open_Session_OK ()
signal Open_Session_REF ()
signal Cancel_Open_Session ()
signal DataAp (Payload : Charstring)
signal Data_Sent ()
signal Data_Error ()
signal Close_Session ()

Figure 5.22. *Simple transport protocol. Modified interface for the chat application*

5.5. Architecture design

5.5.1. *Simple transport protocol*

We can now organize the classes composing the STP, starting with the ADUM.

In Figure 5.23 we can see that we added three ports to the ADUM module: P_ADUM_APP to communicate with the application, P_ADUM_ABPS to communicate with the Alternating Bit Protocol Sender, and P_ADUM_ABPR to communicate with the Alternating Bit Protocol Receiver.

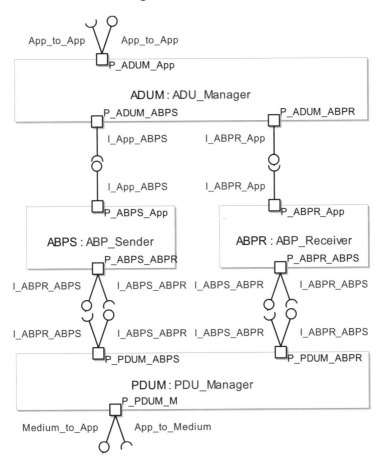

Figure 5.23. *Simple transport protocol. Ports and interfaces for the classes composing the simple transport protocol*

Observe the interfaces associated with each port. Note, for example, that the P_ADUM_ABPS port has no input interface since the ABPS is not supposed to send any information to the ADUM. Conversely, the P_ADUM_ABPR port has no output interface since it is not supposed to send any information through this port.

Now, let us take a look at the ABPS module. We added two ports to this class: P_ABPS_App to communicate with the ADUM module, and P_ABPS_ABPR in order to manage the communication with ABPR. Note that we will connect the P_ABPS_ABPR port to the PDUM module. Also note that this port sends E packets and receives Ack packets.

We can now analyze the ABPR module. We added two ports to this class: P_ABPR_App to send information to the ADUM module, and P_ABPR_ABPS to manage communication with the sender. Note that the P_ABPR_ABPS port receives E packets and sends Ack packets.

Finally, we can analyze the PDUM module. We added three ports to this class: P_PDUM_ABPS to communicate with the ABPS module, P_PDUM_ABPR to communicate with the receiver, and P_PDUM_M to communicate with the medium. Note that this class sends and receives PDUs to and from the medium.

Now, we can connect all the ports, between them and to the external world. Figure 5.24 shows how these classes are structured inside the Simple Transport Protocol module.

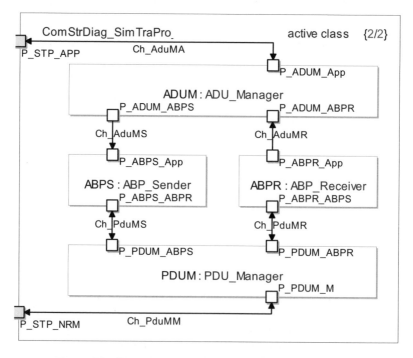

Figure 5.24. *Simple transport protocol. Connecting the modules composing the simple transport protocol*

The external square represents the STP class borders. Note that we added two ports to the STP class: P_STP_APP to communicate with the application and P_STP_NRM to communicate with the Non-Reliable Medium.

Observe that the Ch_AduMS and Ch_AduMR channels are unidirectional and that they connect the ADUM module to ABPS and ABPR.

Finally, let us remark that the Ch_PduMM channel connects the PDUM module to the external world. This channel transports the packets going to and coming from the Non-Reliable Medium.

5.5.2. Simulation model

In order to build and simulate the resulting global model, we need to connect the STP module to the chat application and the Non-Reliable Medium.

Figure 5.25 shows the architecture that we will use for simulating and validating the Simple Transport Protocol.

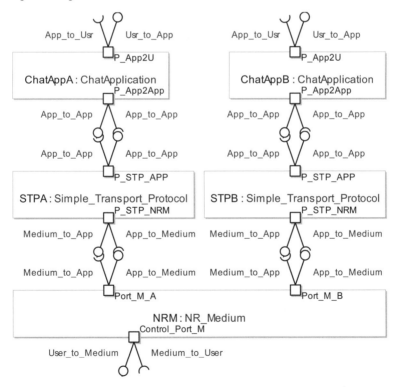

Figure 5.25. *Simple transport protocol. Ports and interfaces for simulating the STP*

In particular, it describes the ports and the interfaces assigned to each class.

Observe that we created two instances of the ChatApplication class, ChatAppA and ChatAppB, and also note the coherence in the interfaces used to communicate with the STP instances.

We should note that we have also created two different instances of the Simple_Transport_Protocols class, STPA and STPB. Each instance is used by a different instance of the ChatApplication class.

Figure 5.26 shows how the modules involved in the simulation are interconnected.

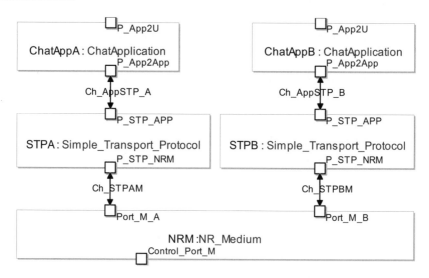

Figure 5.26. *Simple transport protocol. Architecture used to simulate the STP*

5.6. Detailed design

5.6.1. *ADU data type*

The ADU data type is composed of two char strings: the Header and the Payload. The Header contains the name of the message represented by the ADU. The Payload contains the payload carried by the data packet.

We now need to define a basic constructor[1] creating an empty ADU, and also define an empty ADU as ADU("", "").

We also need to define two more constructors: ADU(ADU) which will create a new ADU by copying the parameters of the ADU received; and ADU(header, payload) which will create a new ADU with the received parameters.

Finally, we must define two operations retrieving the internal attributes (commonly called getters): getHeader() and getPayload().

Figure 5.27 shows the detailed ADU class with its attributes and operations.

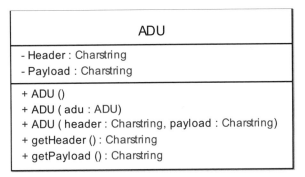

Figure 5.27. *Simple transport protocol. Detailed ADU class with attributes and operations*

5.6.2. *PDU data type*

The PDU data type is composed of three attributes: Header, control bit (CB) and Payload. Remember that the payload in a PDU is of type ADU.

We must define three getters in order to retrieve the private attributes: getHeader(), getCB(), and getPayload().

We also need to define two creators: PDU(header, cb, payload) which will create a new PDU based on the received parameters; and PDU() will create an empty PDU. We can define an empty PDU as PDU("", 999, EMPTY_ADU).

Figure 5.28 shows the detailed PDU class with its attributes and operations.

---

1 Remember that a constructor in a class is a special type of subroutine called at the instantiation of an object. It prepares the new object for use, often accepting parameters which the constructor uses to set any attributes required when the object is first created.

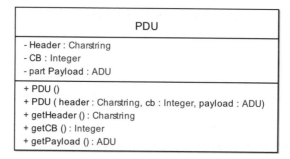

PDU
- Header : Charstring - CB : Integer - part Payload : ADU
+ PDU () + PDU (header : Charstring, cb : Integer, payload : ADU) + getHeader () : Charstring + getCB () : Integer + getPayload () : ADU

Figure 5.28. *Simple transport protocol. Detailed PDU class with attributes and operations*

5.6.3. *ADU_Manager*

Most of the behavior of the ADU_Manager consists of translating every message received from the application into an ADU. The ADUM translates every message into a string of characters and assigns it to the header attribute in the ADU. Then, it assigns the payload (if any) to the Payload attribute.

The state machines for the ADU_Manager use three auxiliary variables: Payload, adu and header (see note (1) in Figure 5.29).

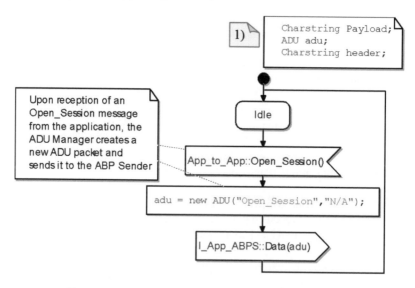

Figure 5.29. *Simple transport protocol. ADUM: translating the Open_Session packet into an ADU*

The ADU_Manager starts in an idle state. On receipt of an Open_Session message from the Chat_Application, it creates a new ADU with an empty payload (represented by the string "n/A"). Then, it sends a Data(ADU) packet to the Alternating Bit Protocol Sender.

Finally, it comes back to the idle state and waits for a new stimulus from either the chat application or the receiver.

Figure 5.30 shows the state machine translating the Data packet received from the Chat_Application. On receipt of this packet, it creates a new ADU with the received payload. Then, it sends the new ADU in a Data packet to the ABPS.

Finally, it comes back to the idle state and waits for a new packet to translate.

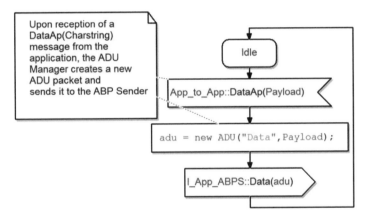

Figure 5.30. *Simple transport protocol. ADUM:*
translating the Data packet into an ADU

The rest of the packets coming from the Chat_Application are translated using the same structure.

The state machines representing these translations are presented in section A.1 in the Appendix.

Figure 5.31 shows the state machine translating the packets coming from the Alternating Bit Protocol Receiver.

On receipt of a data packet, the system reads the header (see note (1)). After that, the ADUM translates the received header into the corresponding packet and sends it to the Chat_Application. Note that the ADUM additionally reads the payload (see note (2)) when the received header is Data.

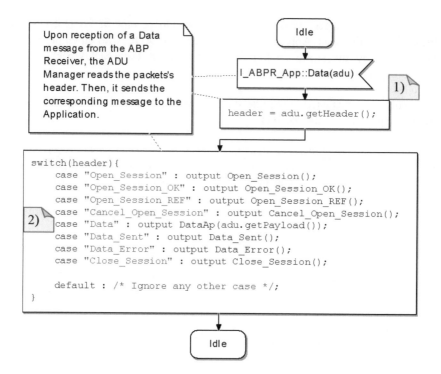

Figure 5.31. *Simple transport protocol. ADUM:*
translating ADU into original packets

5.6.4. PDU_Manager

The work of the PDU manager consists of translating all the messages coming from and going to the ABPS and ABPR into PDUs and vice versa. The PDUM translates every message into a string of characters and assigns it to the Header attribute in the PDU. After that, it adds the Control Bit (CB) and the payload and sends it to the medium.

In the opposite direction, the PDUM extracts all the fields from a PDU and creates simple messages; then, it sends the packet to either the ABPS or ABPR.

The state machines of the PDUM use four auxiliary variables: CB, pdu, header and adu (see note (1) in Figure 5.32).

Figure 5.32 shows the translation of an E packet into a PDU. On receipt of an E packet, the PDUM creates a PDU with the received values. After that, it sends the new PDU to the Non-Reliable Medium. Finally, it comes back to its idle state and waits for a new message to translate. Note that the PDUM does not analyze the received CB or ADU; instead, it only uses them to construct a new PDU.

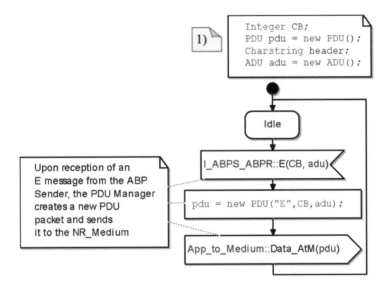

Figure 5.32. *Simple transport protocol. Translation of an E packet into a PDU*

Figure 5.33 shows the translation of an Ack packet into a PDU. On receipt of an Ack packet, the PDUM creates a new PDU with en empty ADU as its payload. After that, it sends the new PDU to the Non-Reliable Medium.

Finally, it returns to its idle state and waits for a new message to translate.

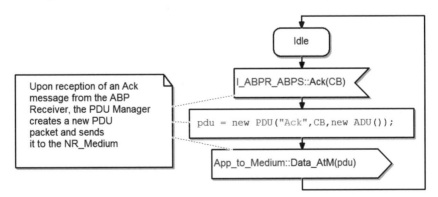

Figure 5.33. *Simple transport protocol. Translation of an Ack packet into a PDU*

Figure 5.34 shows the state machine translating a PDU into an E or Ack packet. On receipt of a data packet from the medium, the PDUM reads the header. It then creates the correct packet and sends it to the ABPS or ABPR modules.

Finally, it goes back to the idle state and waits for another packet to translate.

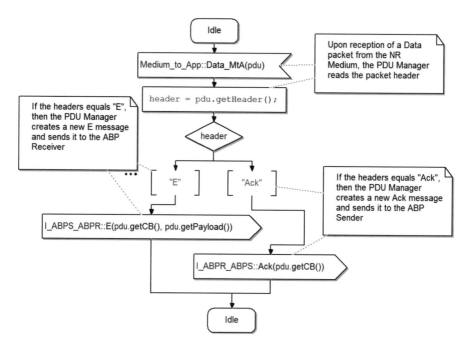

Figure 5.34. *Simple Transport Protocol. Translation of a PDU into an E or Ack packet*

5.6.5. *ABP_Sender*

The ABP_Sender class uses an internal attribute to define the time it will wait until re-sending the E packet when no answer is received from the ABPR. We can call this attribute Maximum Round Trip Time (RTT).

The sender also needs a timer. We will call this timer timerABPS.

Figure 5.35 shows the detailed ABP_Sender class with attributes and operations.

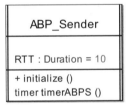

Figure 5.35. *Simple Transport Protocol. Detailed ABP_Sender class with attributes and operations*

Figure 5.36 shows the behavior of the ABPS when it is waiting for a first data to send. The system starts in the St_WaitingData_0 state. On receipt of a data packet from the ADUM, it sends an E packet to the PDUM with CB equal to 0 and the received ADU (observe the initialization of the oldCB variable). After that, the ABPS sets its timer to now + RTT. Finally, it goes to the St_WaitingAck_0 state and waits for an acknowledgement from the Alternating Bit Protocol Receiver.

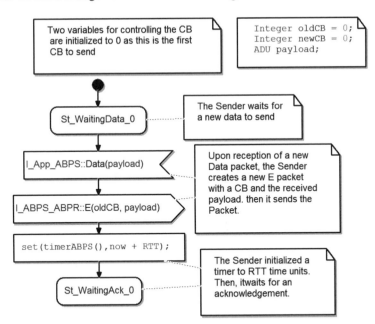

Figure 5.36. *Simple transport protocol ABP_Sender class waiting for the first data to send*

If the ABPS receives an unexpected acknowledgement while waiting for a new data, it simply ignores it. This behavior is represented in Figure 5.37.

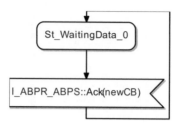

Figure 5.37. *Simple transport protocol. ABP_Sender class ignoring unexpected acknowledgements while waiting for new data*

Figure 5.38 shows the behavior of the ABPS class when it is in the St_WaitingAck_0 state and receives an Ack packet from the ABPR.

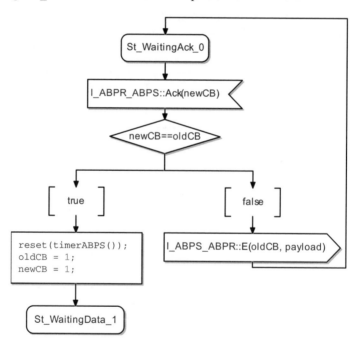

Figure 5.38. *Simple transport protocol. ABP_Sender class waiting for the Ack(0) from the ABPR*

On receipt of an Ack packet, the ABPS reads the received CB. Then, it compares it with the last CB sent.

If the received CB is equal to the expected one, then the ABPS resets its timer. After that, it updates its internal CB variables and waits for a new data to send.

If the received CB is not equal to the expected one, then the ABPS re-sends the E packet to the PDUM and goes back to waiting for the expected acknowledgement.

If the ABPS is waiting for an Ack(0) from the ABPR and the internal timeout signal is triggered, then the ABPS re-sends the E packet to the PDUM and sets its timer again. After that, it goes back to waiting for a correct answer from the ABPR. Figure 5.39 shows the state machine modeling this behavior.

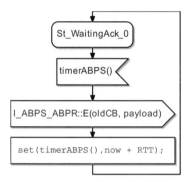

Figure 5.39. *Simple transport protocol. ABP_Sender class: timeout when waiting for Ack(0)*

After reception of the acknowledgement of the E packet with CB 0, the ABPS waits for a new data to send. When it receives a new data packet from the ADUM, the ABPS sends an E packet to the PDUM with CB equal to 1 and the received payload. After that, it sets its timer to now + RTT. Finally, it waits for an acknowledgement from the ABPR. Figure 5.40 shows the state machine modeling this behavior.

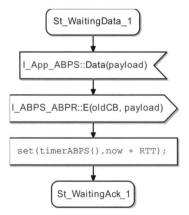

Figure 5.40. *Simple transport protocol. ABP_Sender class waiting for the second data to send*

If the ABPS receives an unexpected acknowledgement while waiting for a new data, it simply ignores it. This behavior is represented in Figure 5.41.

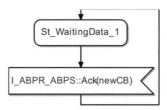

Figure 5.41. *Simple transport protocol. ABP_Sender class ignoring unexpected acknowledgements while waiting for new data*

When the ABPS is in the St_WaitingAck_1 state and it receives an Ack packet, it reads the received CB and compares with the last CB sent. If this is equal to the one expected, it resets its timer and updates its internal variables. After that, it goes to the initial state. If the received CB is not equal to the expected one, then it re-sends the E packet to the PDUM and goes back to waiting for the correct acknowledge packet. Figure 5.42 shows the state machine implementing this behavior.

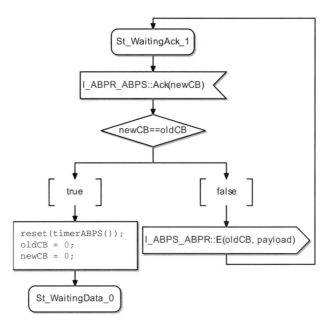

Figure 5.42. *Simple transport protocol. ABP_Sender class waiting for the Ack(1) from ABPR*

After RTT time units, if no correct Ack packet is received, the internal timeout signal is triggered. After that, the ABPS re-sends the E packet to the PDUM. Then, it sets its timer again. Finally, it goes back to waiting for the correct Ack packet from the ABPR. The state machine in Figure 5.43 describes this behavior.

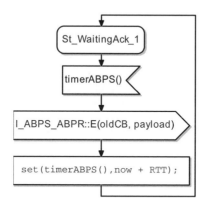

Figure 5.43. *Simple transport protocol. ABP_Sender class: timeout when waiting for Ack(1)*

5.6.6. *ABP_Receiver*

The Alternating Bit Protocol Receiver uses three internal variables: oldCB and newCB for controlling the received CB; and adu for reading the received payload (see Figure 5.44).

At the start, the ABPR waits for an E packet with a CB equal to 0. On receipt of an E packet from the ABPS, it acknowledges it with the received CB. After that, it compares the newly received CB with the last one received.

If the received CB is equal to the last received, it means that the ABPR has already processed the received payload; it then ignores the payload and goes back to waiting for an E packet.

If the newly received CB is different from the last received one, it means that this is a new packet and the ABPR will process the received payload. Then, it sends a data packet containing the received payload to the ADUM. Finally, it updates its control variables and goes to the St_WaitingE_1 state and waits for a new packet from the sender.

Figure 5.44 shows the state machine modeling this behavior.

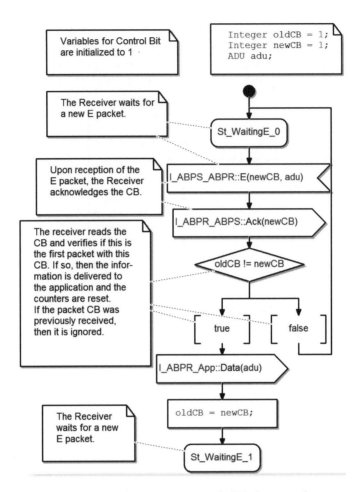

Figure 5.44. *Simple transport protocol. ABP_Receiver class:*
waiting for an E packet with CB equal to 0

Figure 5.45 shows a state machine modeling the behavior of the ABPR when it waits for an E packet containing a CB equal to 1.

On receipt of an E packet, it acknowledges the received CB; then it compares the last received CB against the new one in order to know if the E packet contains new data or not. If the new CB and the last received CB are equal, it means that the received payload has already been processed. Furthermore, it means that the received payload has been forwarded to its user. In that case, the ABPR sends a data packet to the ADUM and updates its internal controller variables. Finally, it goes to the St_WaitingE_0 state waiting for a new data.

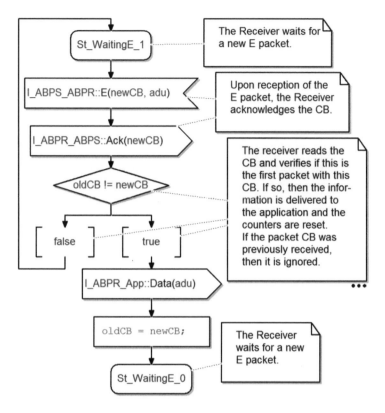

Figure 5.45. *Simple transport protocol. ABP_Receiver class:*
waiting for an E packet with CB equal to 1

5.7. Simulations

The sequence diagrams resulting from the simulations are quite large since our simulation software creates a lifeline for every object in the model. For this reason, in this section, we will mainly only show extracts of the main simulation diagram.

The detailed and extended diagrams are presented in the Appendix.

5.7.1. *Initialization and configuration*

First of all, we need to verify that all the objects involved in our simulations are correctly initiated at the beginning of each execution. This means that we must verify that all objects are in their idle state waiting for an external stimulus.

Figures 5.46 and 5.47 show all objects in their initial state.

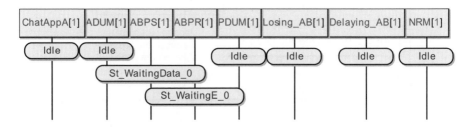

Figure 5.46. *Simple transport protocol. Simulation:*
all the objects are in their initial state (part 1)

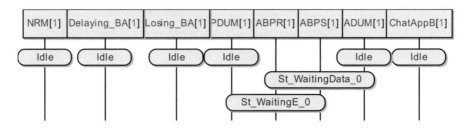

Figure 5.47. *Simple transport protocol. Simulation:*
all the objects are in their initial state (part 2)

Now we can verify that we can correctly access the NR_Medium and that we can configure it.

Figure 5.48 shows the sequence diagram resulting from the simulation of the medium configuration. Observe that the simulation corresponds to the results obtained in section 4.6.1.

The sequence diagram presented here is only an excerpt of the entire diagram, and focuses on our target simulation and the objects concerned. The detailed and extended diagrams are shown in Figure A.7 in the Appendix.

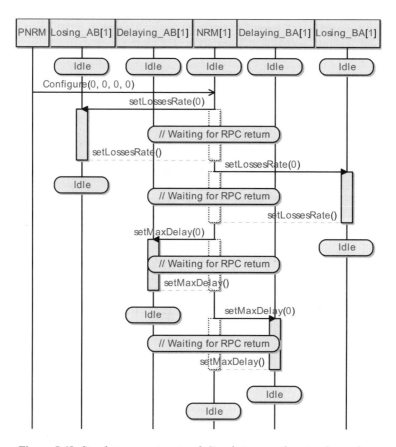

Figure 5.48. *Simple transport protocol. Simulation: configuring the medium*

5.7.2. *No losses*

Let us verify some of the already known scenarios. In this section, we will test how the simulation model reacts when there are no losses in the medium. For this, we will execute an open session request from User A without an answer from User B. Then we will again simulate the same scenario but, this second time, User B will accept the request.

5.7.2.1. *Open session request without answer*

Note the first part of the sequence diagram obtained from simulation in Figure 5.49. The diagram shows the internal behavior of the Non-Reliable Medium (see note (1)), i.e. the packets exchanged between Losing_AB, Delaying_AB, Losing_BA and Delaying_BA classes. However, we tested this behavior in the last chapter and we are not interested in testing it again. Moreover, these lifelines might enlarge the resulting diagrams and complicate the reading.

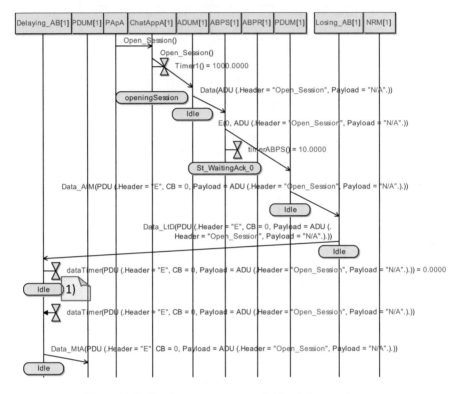

Figure 5.49. *Simple transport protocol. Simulation: no losses, open session req without answer (detailed)*

We can run the simulation again. This time, we define the Non-Reliable Medium as a black-box, which means that we will not be able to see any of the interactions inside the NRM object. This new simulation allows us to focus on the protocol behavior.

Figure 5.50 shows the first part of this new simulation. Observe in note (1) that the chat application sets its internal timer to 1,000 time-units. Note also that the ADUM encapsulates the received packet and sends it forward to the ABPS (see note (2)). After that, the ABPS creates an E packet containing the received ADU; then, it sets its internal timer to 10 time-units. We can also note that the PDUM correctly creates a PDU describing the received packet (containing a header, a CB and an ADU) and forwards it to the NRM.

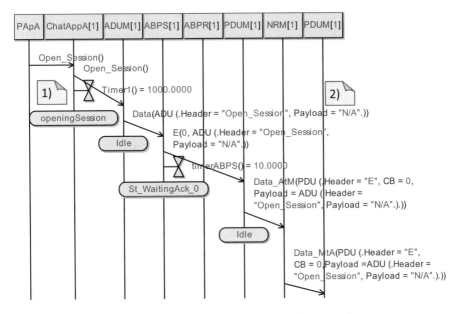

Figure 5.50. *Simple transport protocol. Simulation: no losses,*
open session req without answer (Simplified 01)

Figure 5.51 shows the continuation of the scenario. Observe, in note (1), that PDUM_B extracts the PDU from the received packet and forwards an E Packet to ABPR_B. ABPR_B acknowledges the received packet and forwards the received payload to the ADUM (see note (2)). Note also that the Open_Session packet correctly arrives to User B (represented here by PApB). Finally, note that the acknowledgement containing a CB equal to 0 correctly goes back and is received by ABPS_A (see note (3)).

We can now describe the state of the system at this point:

– first, ABPS_A knows that the packet was correctly received and it is waiting for a new data to transmit. It is in its St_WaitingData_1 state and has reset its internal timer;

– second, ChatAppB waits for an answer from its user and is in its waitingForAnswer state; and

– third, ChatAppA is in its openingSession state and waits for an answer from ChatAppB.

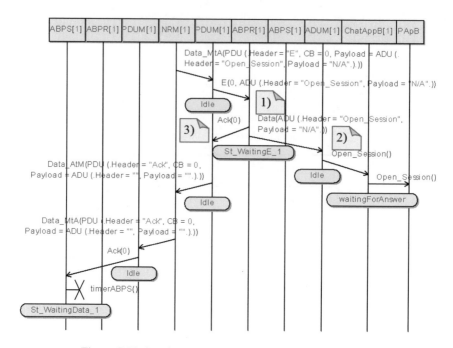

Figure 5.51. *Simple transport protocol. Simulation: no losses,*
open session req without answer (Simplified 02)

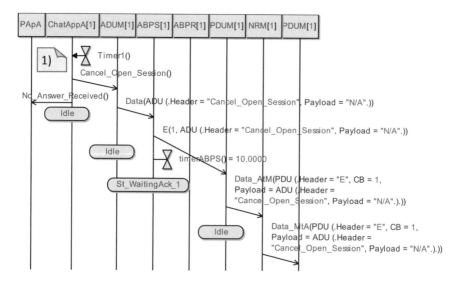

Figure 5.52. *Simple transport protocol. Simulation: no losses,*
open session req without answer (Simplified 03)

Figure 5.52 shows the continuation of this simulation.

We can observe in note (1) that ChatAppA's internal timer is over. It then sends a Cancel_Open_Session and informs its user. The application packet is translated into an ADU and then into a PDU, and then sent to the medium.

Note in Figure 5.53 that the PDU is converted back into an ADU and into an application packet, and then delivered to ChatAppB. We can observe that a protocol-level acknowledgement containing a CB equal to 1 is sent back to ABPS_A. On receipt of this acknowledgement, ABPS_A resets its internal timer and goes to the St_WaitingData_0 state to wait for a new data to send.

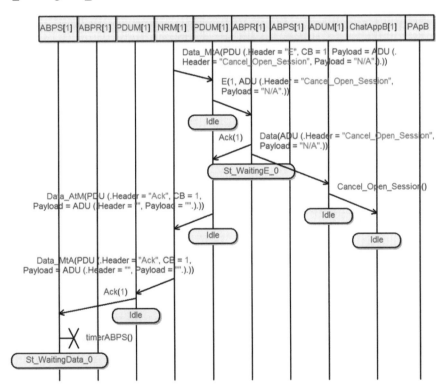

Figure 5.53. *Simple transport protocol. Simulation: no losses, open session req without answer (Simplified 04)*

A detailed diagram of this simulation can be found in Figures A.8 and A.9 in the Appendix.

5.7.2.2. *Open session request with answer*

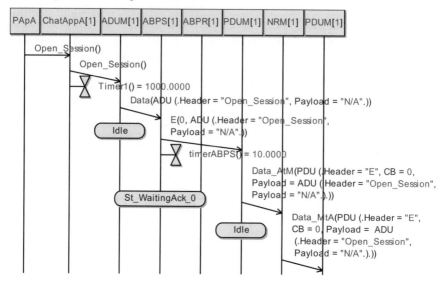

Figure 5.54. *Simple transport protocol. Simulation: no losses,*
open session req with answer (01)

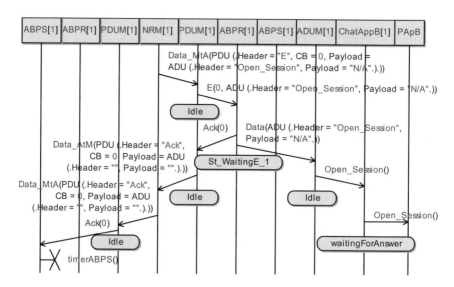

Figure 5.55. *Simple transport protocol. Simulation: no losses,*
open session req with answer (02)

In Figure 5.54, we see that User A (represented here by PApA) sends a second Open_Session request. The sequence of messages shown in this diagram is the same as that in the previous simulation.

Then, in Figure 5.55 we can observe how the request is delivered to User B (represented here by PApB) and how the protocol-level acknowledgement is sent back to ABPS_A.

After this, User B accepts the session request, as shown in note (1) in Figure 5.56. In this figure, we can observe how the request is converted into an ADU, then into a PDU, then sent to the medium and finally delivered to the PDUM_A.

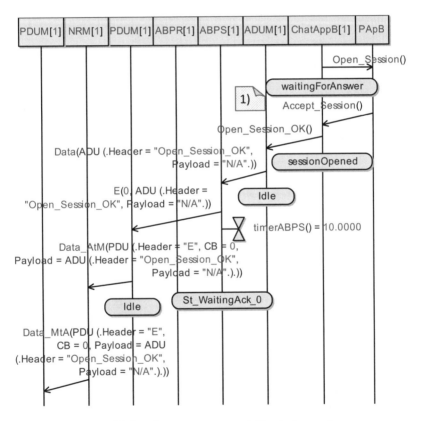

Figure 5.56. *Simple transport protocol. Simulation: no losses,*
open session req with answer (03)

Figure 5.57 shows the session acknowledgement being delivered to ChatAppA. It also shows the protocol acknowledgement sent back by ABPR_A.

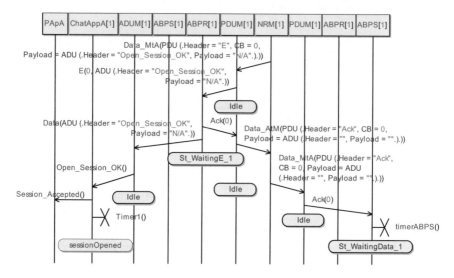

Figure 5.57. *Simple transport protocol. Simulation: no losses, open session req with answer (04)*

A detailed diagram of this simulation can be seen in Figure A.10 in the Appendix.

5.7.2.3. *Send data*

Let us suppose that the session is already established between the applications. Now, let us send a message from User A to User B and check what happens. Figure 5.58 shows the first part of the sequence diagram resulting from this simulation.

First, we can observe that all the objects in the diagram are in a stable state, i.e. no object is waiting for an answer.

Observe in note (1) (Figure 5.58) that User A (represented here by PApA) sends "Hello world" to his application. ChatAppA sends the received payload into the DataAP packet to the ADUM_A.

We can see that the packet is encapsulated into an ADU and then into a PDU. We can also observe that ABPS sets its internal timer to 10 time-units.

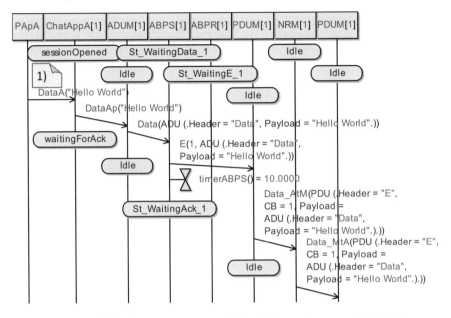

Figure 5.58. *Simple transport protocol. Simulation: no losses, send data (01)*

We can see in Figure 5.59 that the user message is correctly delivered to its destination and the protocol acknowledgement is sent back to ABPS.

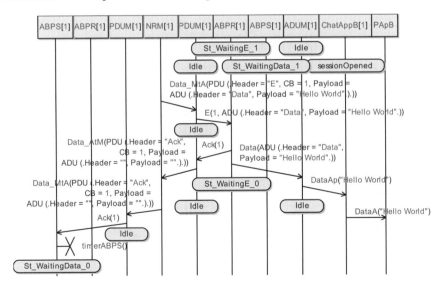

Figure 5.59. *Simple transport protocol. Simulation: no losses, send data (02)*

Figure 5.60 shows that ChatAppB delivers the user message to its destination and then sends a Data_Sent packet towards ChatAppA. We can see, again, how the packet passes through the transport protocol layer, then goes to the NRM and finally to the protocol layer.

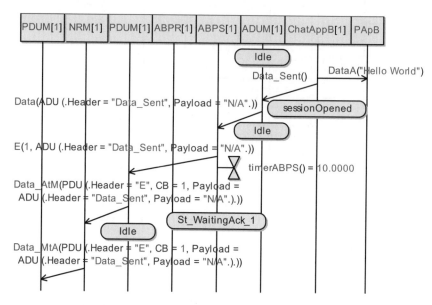

Figure 5.60. *Simple transport protocol. Simulation: no losses, send data (03)*

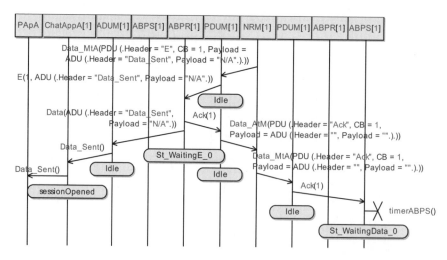

Figure 5.61. *Simple transport protocol. Simulation: no losses, send data (04)*

Finally, observe in Figure 5.61 how the application acknowledgements cross the protocol layer and then the application. Note that User A is informed by his application about the successful data transmission.

Finally, notice how the protocol acknowledgement is sent back and how the sender resets its internal timer.

A diagram containing the entire and detailed simulation of this scenario appears in Figure A.11 in the Appendix.

5.7.3. *Losses*

We can now test the recovery mechanisms implemented by the ABP. To do so, we will define a low reliability channel from A to B and a full reliability one from B to A.

We can test these conditions for a data transmission process. Remember that a data transmission from A to B is divided into two parts. In the first part, ChatAppA sends a data packet to B, while in the second part, ChatAppB answers with a Data_Sent packet. In the first part of the scenario we expect that, at protocol level, the E packet going from A to B will be lost by the medium and that the ABP modules will have to recover it. In the second part of the scenario, we expect that the protocol acknowledgement will be lost by the medium

For the moment we will not define any delay; however, in the next section, we will perform a simulation including some delay in the medium in order to illustrate how this ABP behaves under these circumstances.

Figure 5.62 shows the first part of this scenario. First, User A sends a data packet to his application, and the data packet contains the message "Hello World". The Data packet is received by the application. Then it is forwarded to ADUM_A. After that, the application waits for an acknowledgement from ChatAppB.

The data packet arrives to ABPS_A. This module sends an E packet to PDUM_A; then it sets its internal timer and waits for an acknowledgement from ABPR_B.

Observe in note (1) that the medium loses the data packet. After some time, ABPS_A resends the data packet (see note (2)) and the medium loses the packet again. In the simulation that we perform, the packet is lost three times by the medium. However we will not show all the messages performing the resending operation in this section.

The entire and detailed sequence diagram is given in Figure A.12, Appendix.

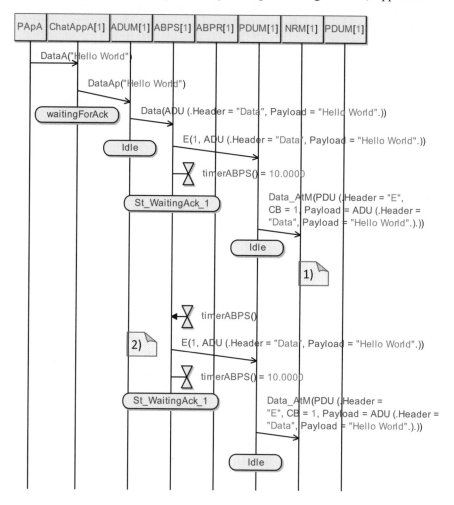

Figure 5.62. *Simple transport protocol. Simulation: losses, send data (01)*

Observe in Figure 5.63 that ABPS resends the data packet and the medium delivers it to the receiver side. After that, the data packet is delivered to ABPR. Then, the message is sent to the chat application while a protocol acknowledgement is sent back to the sender.

Remember that we defined a full reliability from B to A; thus, the acknowledgement is delivered without problems, as expected. Note that, on receipt of the protocol acknowledgement, ABPS_A resets its internal timer and goes to the St_WaitingData_0 state and waits for a new data to transmit.

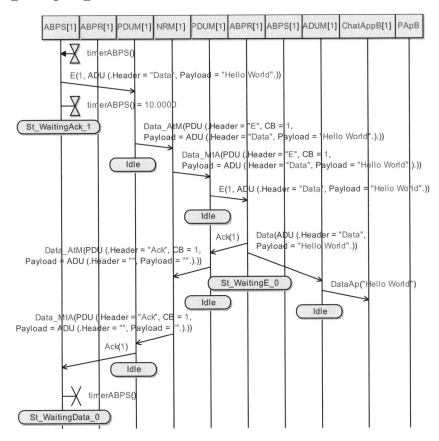

Figure 5.63. *Simple transport protocol. Simulation: losses, send data (02)*

Figure 5.64 shows the instant when the data packet is delivered to ChatAppB. On receipt of the data packet, ChatAppB delivers it to its user and sends an acknowledgement packet to the sender application. Observe in note (1) that the packet correctly reaches the medium and then the PDUM_A.

In Figure 5.65 we can observe that the packet is delivered to ABPR_A. On receipt of the packet, ABPR_A sends an acknowledgement back to ABPS_B and sends the application acknowledgement towards ChatAppA. Observe in note (1) that the application receives the acknowledgement and informs its user. Note also that the protocol acknowledgement correctly reaches the medium, but it is not retransmitted (see note (2)).

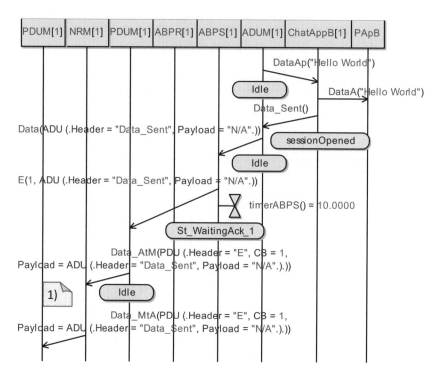

Figure 5.64. *Simple transport protocol. Simulation: losses, send data (03)*

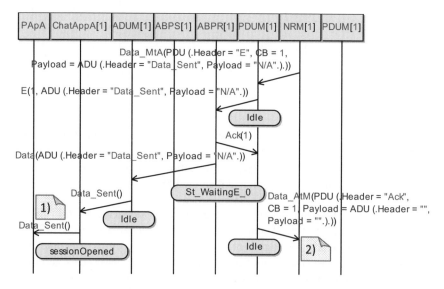

Figure 5.65. *Simple transport protocol. Simulation: losses, send data (04)*

Figure 5.66 shows the moment where ABPS_B receives a timeout. On receipt of this signal, ABPS_B resends the data packet and waits for the protocol level acknowledgement.

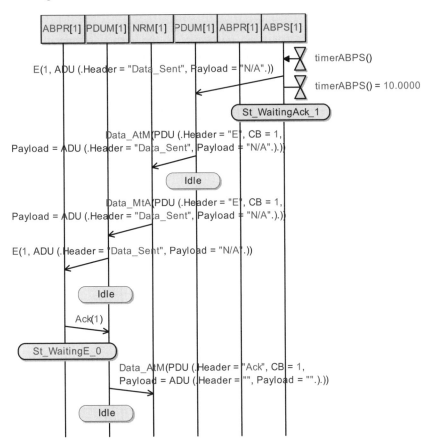

Figure 5.66. *Simple transport protocol. Simulation: losses, send data (05)*

Again, the data packet is received by ABPR_A. This time, ABPR_A does not forward the packet towards the application since the received CB indicates that the packet was already received. Note that ABPR_A is waiting for a CB equal to 0, but the received CB is 1. Note also that the protocol acknowledgement correctly arrives at the medium, but it is not forwarded.

In this simulation, the protocol level acknowledgement was lost four times by the medium.

Not all the messages performing the resending operation are shown in this section. However, the entire and detailed sequence diagrams can be seen in Figures A.13 and A.14 in the Appendix.

Finally, Figure 5.67 shows that the protocol acknowledgement successfully goes back to *ABPS_B*. Again, note that, on receipt of the acknowledgement, *ABPS_B* resets its internal timer, and waits for a new data to transmit.

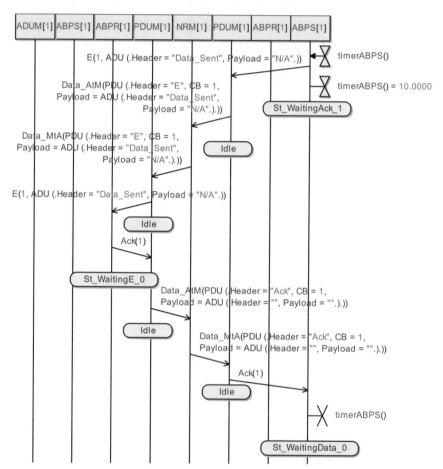

Figure 5.67. *Simple transport protocol. Simulation: losses, send data (06)*

This scenario allowed us to verify that our implementation of the Alternating Bit Protocol correctly transmits data packets over an unreliable network. It shows how the sender resends the data packets at regular intervals and how it manages the acknowledgements at different moments. It also shows how the receiver manages the repeated packets. It finally illustrates how the protocol behaves when either the E or the Ack packets are lost by the medium.

5.8. Further considerations

The Alternating Bit Protocol provides a reliable transport service on a non-reliable network. However, our current implementation of this algorithm cannot guarantee a reliable service when the network suffers from a high and variable delay. Indeed, the calculation of the timeout assigned to the protocol sender should mandatorily be set at least to the maximum round trip time; however, in our case, we set this timeout arbitrarily to 10 time units regardless of the maximum round trip time.

Let us now analyze two possible scenarios and observe how the protocol reacts under these circumstances.

5.8.1. *Acknowledgements at different levels*

The Simple Transport Protocol that we have modeled provides a reliable full duplex communication service to an application. This protocol implements the full duplex communication by duplicating the Alternating Bit unidirectional channel. As a consequence of this duplication, there is a separated channel for communicating from one point to another. This double channel at the protocol level may have an unexpected effect on applications. Indeed, the application-level data packets and the application-level acknowledgements travel on separate channels, and are then potentially submitted to different reliability and delay rates. Thus, an application level acknowledgement may arrive before a protocol level one.

We can imagine the following scenario:

– user A sends "Message 1" to his application;

– application A sends a data packet to Protocol A;

– protocol A sends an E packet to Protocol B;

– protocol B receives the data packet, then acknowledges the packet and forwards the payload to Application B;

– the acknowledgement sent by Protocol B is highly delayed by the medium;

– application B receives the data packet, then acknowledges the packet and forwards the payload to User B;

– the application-level acknowledgement sent by Application B arrives with no delay to Application A;

– user A sends "Message 2" to his application;

– application A sends a data packet to Protocol B;

– protocol B cannot receive a new data because it is still waiting for the delayed protocol-level acknowledgement. Thus, the data packet is lost;

– application A is blocked waiting for an acknowledgement from Application B.

Figure 5.68 briefly illustrates this scenario.

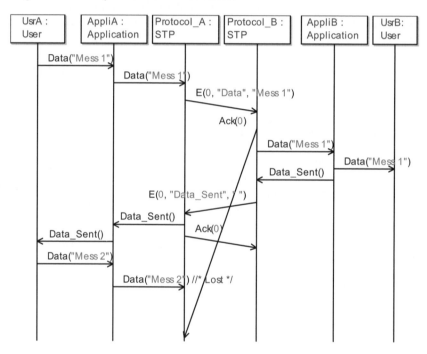

Figure 5.68. *Simple transport protocol. Acknowledgement at different levels*

It is possible to validate this scenario by simulation. Section A.2.4 in the Appendix shows a sequence diagram obtained by simulation. Figure A.18 (Appendix) shows the moment at which the data packet sent from application to protocol is lost.

We will not go into more detail about this simulation since the full validation of the protocol is not within the scope of this book. However, it is enough to say that a solution to this problem may consist of adding a buffer to the protocol sender in order to wait until the protocol is ready to process the received message.

5.8.2. *Delayed acknowledgements*

Let us now imagine a situation where a highly delayed packet wrongly acknowledges another one. We can analyze the following scenario:

– Sender sends "message 1" to Receiver (E(CB=0));

– the packet is highly delayed by the network;

– the Sender considers the packets as lost since no answer is received (timeout);

– Sender resends "message 1' to Receiver (CB = 0);

– Receiver receives "message 1" and acknowledges its receipt (CB = 0);

– Sender receives the acknowledgement with CB = 0 and sends "message 2' to Receiver (CB = 1);

– Receiver receives the delayed "message 1" and acknowledges it (CB = 0). Acknowledgement with CB = 0 is highly delayed by the medium;

– Receiver receives "message 2" and acknowledges its receipt (CB = 1);

– Sender receives the acknowledgement with CB = 1 and sends "message 3' to Receiver (CB = 0);

– "message 3" is lost by the network;

– Sender receives the delayed acknowledgement with CB = 0 and thinks that this is the acknowledgement of "message 3" (the last sent packet);

– Sender sends "message 4" to Receiver (CB = 1).

Note that "message 3" was lost by the medium but the sender does not realize it.

Figure 5.69 briefly illustrates this scenario.

The explanation of this problem is that the timeout assigned to the server is much shorter than the maximum delay experienced by a packet. We can solve this problem by incrementing the sender's timeout. Indeed, the timeout assigned to the server must be at least a maximum round trip time.

A different solution to this problem consists of changing the range of the control header. Remember that the control header in the Alternating Bit protocol is called Control Bit, and it can only take values 0 or 1. By defining a larger control header (for example between 0 and 100), the protocol sender can ignore a very old acknowledgement.

It should be noted that several protocols and algorithms correctly deal with this kind of problem.

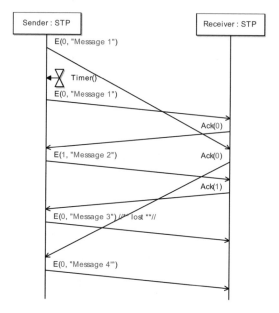

Figure 5.69. *Simple transport protocol. Illustrating the limits of the ABP algorithm*

5.9. Chapter summary

This chapter has concluded the book by linking all the preceding models together in one single model.

We have shown that it is possible to model different communication layers without using a particular programming language. This kind of analysis and design is usually called Model Driven Architecture (MDA).

We have shown in a practical way that MDA is highly enhanced by using Unified Modeling Language (UML).

All the examples presented in this book demonstrate that it is easily possible to model, analyze and test a communicating system by using an object-oriented approach. Of course, these examples do not pretend to undermine the traditional formal methods; on the contrary, our aim is to show some complementary tools which might greatly enhance the software development lifecycle.

The iterative and incremental design method used in this book needs to be underlined. This kind of approach is typically described by the Unified Process (UP). It encourages the division of a huge problem into small blocks. Each block is analyzed, conceived and implemented in a single iteration. Every iteration adds more complexity to the final model.

Finally, note that we used three different tools to create the models presented here. With the use of these tools, we can remark that the modeling language and the modeling process are independent from the software tools.

5.10. Bibliography

[DIA 09] DIAZ M. (ed.), *Petri Nets*, ISTE Ltd., London, John Wiley & Sons, New York, 2009.

Appendix

Detailed Diagrams
of the Simple Transport Protocol

A.1. State machines for the Application Data Unit Manager (Simple Transport Protocol)

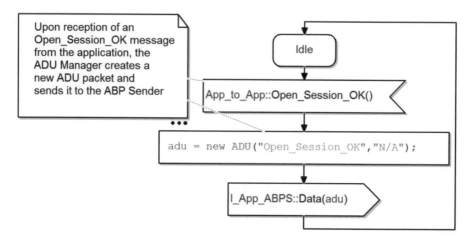

Figure A.1. *Simple transport protocol. ADUM: translating the Open_Session_OK packet into an ADU*

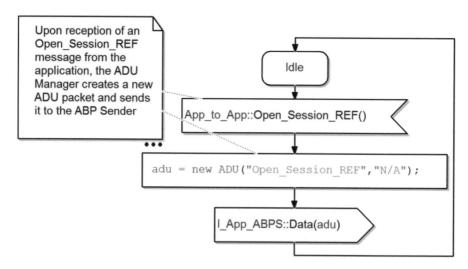

Figure A.2. *Simple transport protocol. ADUM: translating the Open_Session_REF packet into an ADU*

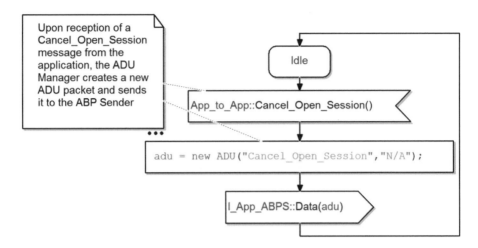

Figure A.3. *Simple transport protocol. ADUM: translating the Cancel_Open_Session packet into an ADU*

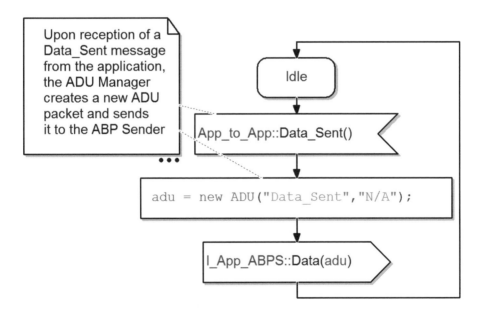

Figure A.4. *Simple transport protocol. ADUM: translating the Data_Sent packet into an ADU*

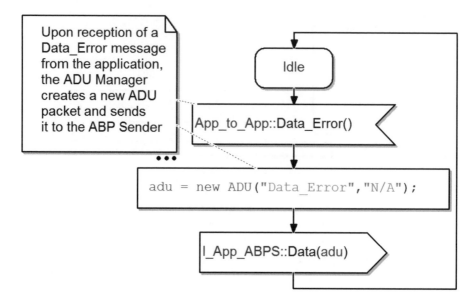

Figure A.5. *Simple transport protocol. ADUM: translating the Data_Error packet into an ADU*

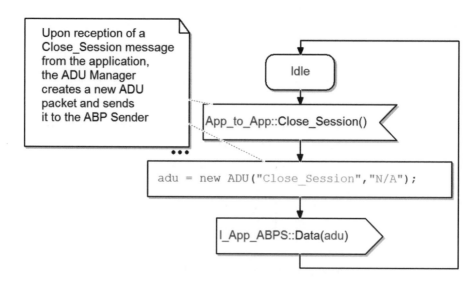

Figure A.6. *Simple transport protocol. ADUM: translating the Close_Session packet into an ADU*

A.2. Detailed simulations of the Simple Transport Protocol

A.2.1. *Initializing and configuring the medium*

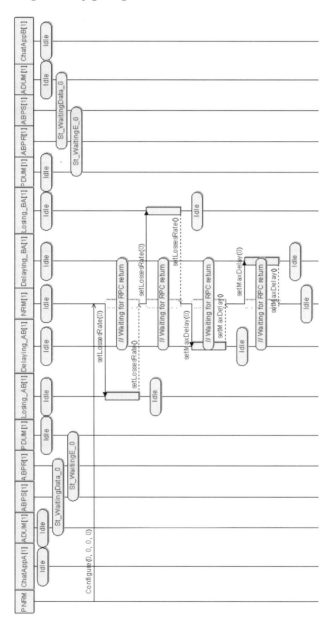

Figure A.7. *Simple transport protocol. Simulation: configuring the medium (detailed)*

A.2.2. *No losses*

A.2.2.1. *Open session request without answer*

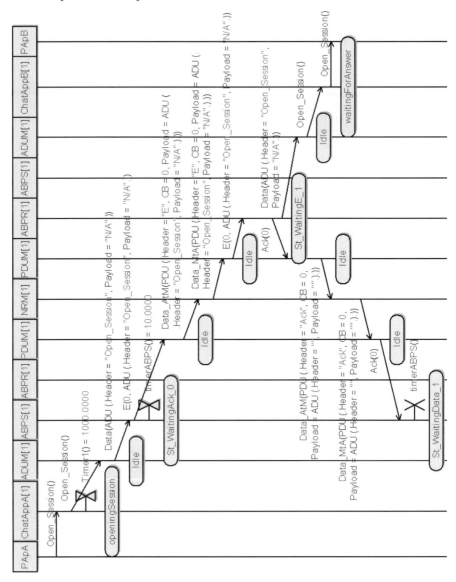

Figure A.8. *Simple transport protocol. Simulation: no losses,*
open session req without answer (Detailed 01)

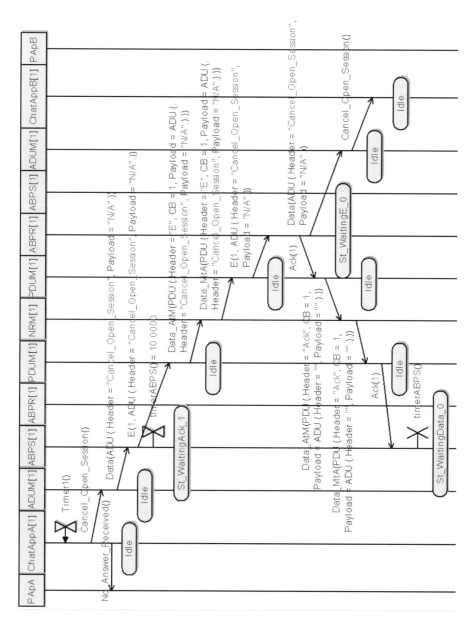

Figure A.9. *Simple transport protocol. Simulation: no losses, open session req without answer (Detailed 02)*

A.2.2.2. *Open session request with answer*

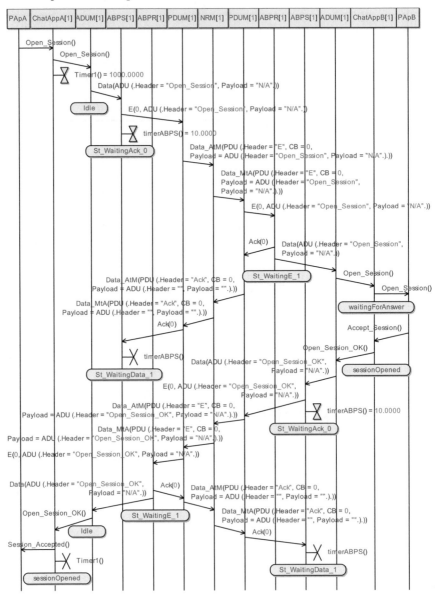

Figure A.10. *Simple transport protocol. Simulation: no losses,
open session req with answer (Detailed)*

A.2.2.3. Send data

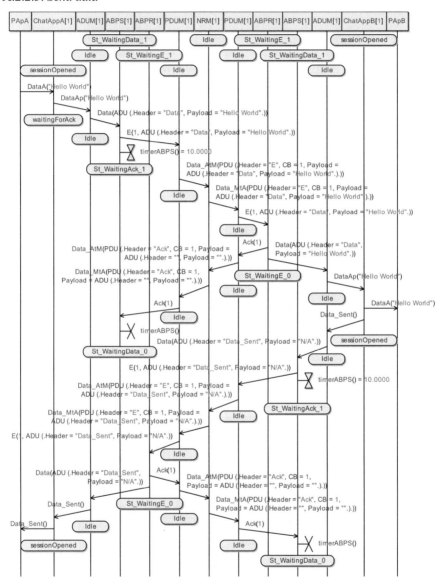

Figure A.11. *Simple transport protocol. Simulation: no losses, send data (detailed)*

A.2.3. *Losses*

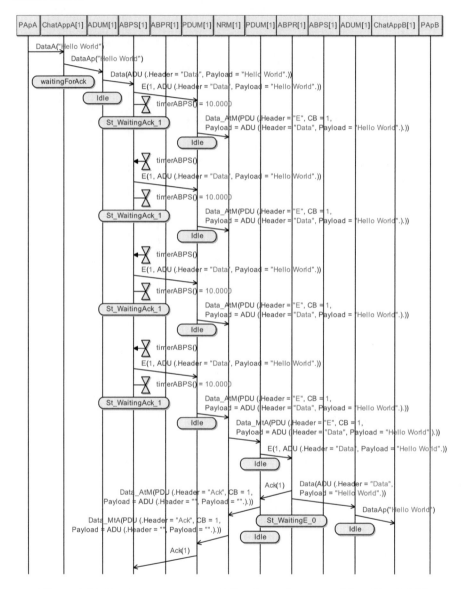

Figure A.12. *Simple transport protocol. Simulation: losses, send data (detailed 01)*

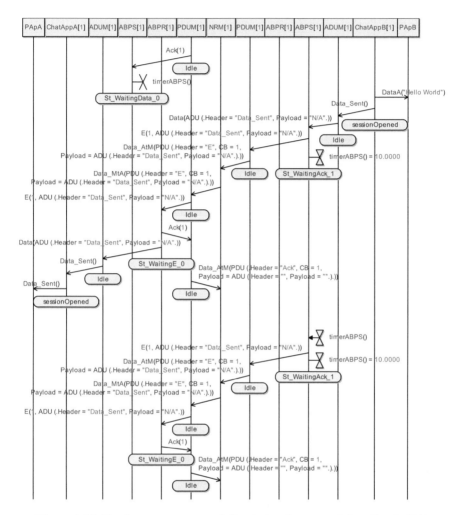

Figure A.13. *Simple transport protocol. Simulation: losses, send data (detailed 02)*

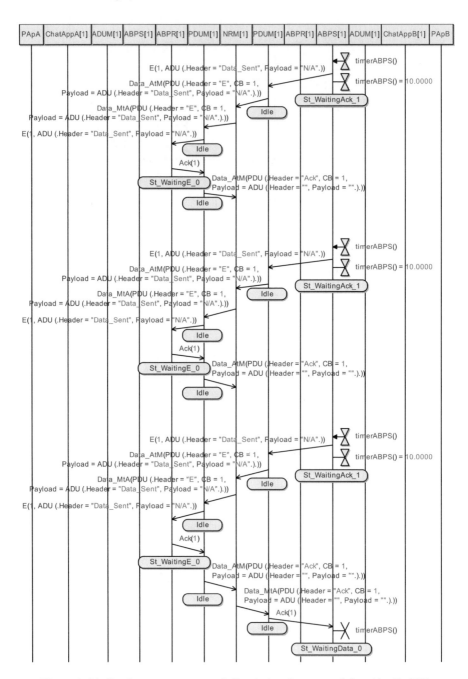

Figure A.14. *Simple transport protocol. Simulation: losses, send data (detailed 03)*

A.2.4. *Losses and delay*

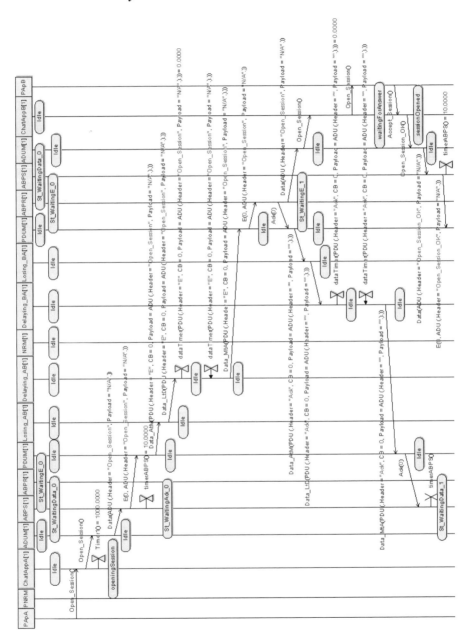

Figure A.15. *Simple transport protocol. Simulation: losses and delay; problems when acknowledging at two levels – opening session (Detailed 01)*

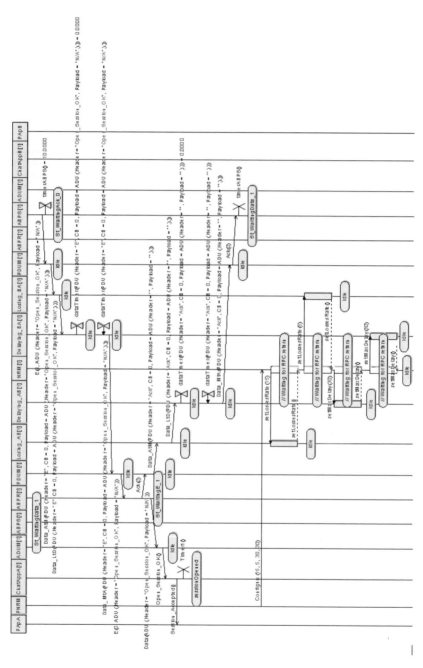

Figure A.16. *Simple transport protocol. Simulation: losses and delay; problems when acknowledging at two levels – end of opening session and configuration (detailed 02)*

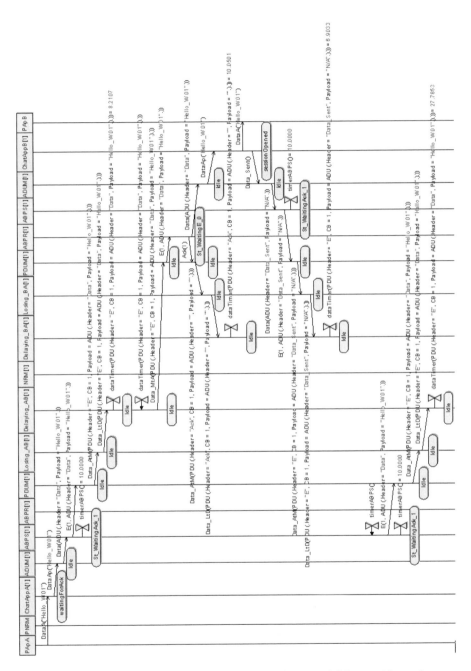

Figure A.17. *Simple transport protocol. Simulation: losses and delay; problems when acknowledging at two levels – send data then acknowledge (Detailed 03)*

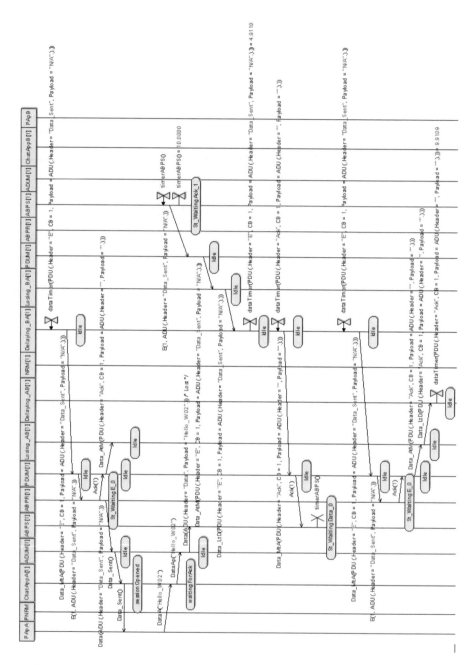

Figure A.18. *Simple transport protocol. Simulation: losses and delay; problems when acknowledging at two levels – discarded packet at protocol level (detailed 04)*

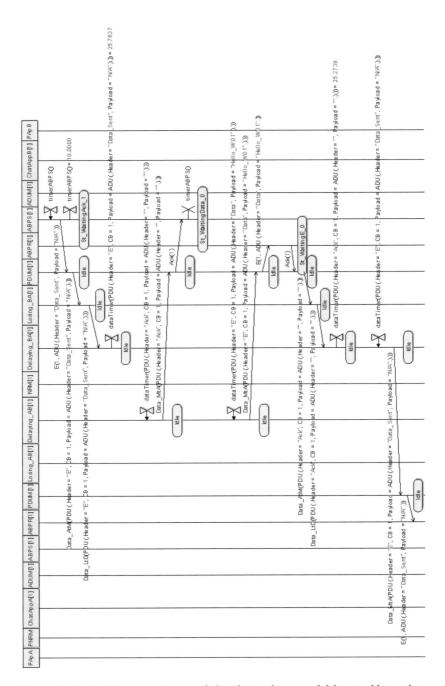

Figure A.19. *Simple transport protocol. Simulation: losses and delay; problems when acknowledging at two levels – resending acknowledgements (detailed 05)*

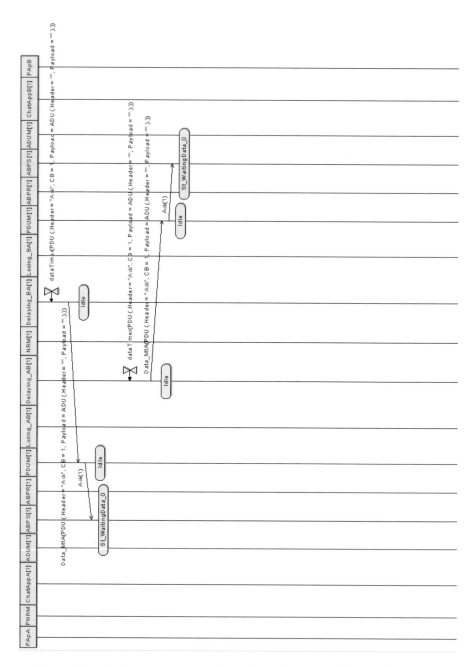

Figure A.20. *Simple transport protocol. Simulation: losses and delay; problems when acknowledging at two levels – all the acknowledgements are received (detailed 06)*

Index